T0186824

Rubber Materials

Rubber Materials

By

B. Kothandaraman

Asst. Prof., Rubber & Plastics Technology
MIT Campus, Anna University,
Chennai-600044

Taylor & Francis
Taylor & Francis Group
Boca Raton   London   New York

CRC is an imprint of the Taylor & Francis Group,
an informa business

Ane Books India

© Ane Books India

First Published in 2008 by

Ane Books India

4821 Parwana Bhawan, 1st Floor
24 Ansari Road, Darya Ganj, New Delhi -110 002, India
Tel: +91 (011) 2327 6843-44, 2324 6385
Fax: +91 (011) 2327 6863
e-mail: anebooks@vsnl.com
Website: www.anebooks.com

For

CRC Press

Taylor & Francis Group
6000 Broken Sound Parkway, NW, Suite 300
Boca Raton, FL 33487 U.S.A.
Tel : 561 998 2541
Fax : 561 997 7249 or 561 998 2559
Web : www.taylorandfrancis.com

For distribution in rest of the world other than the Indian sub-continent

ISBN-10 : 1 42007 159 9
ISBN-13 : 978 1 42007 159 7

All rights reserved. No part of this publication may be reproduced, stored in a retrieval system, or transmitted in any form or by means, electronic, mechanical, photocopying, recording and/or otherwise, without the prior written permission of the publishers. This book may not be lent, resold, hired out or otherwise disposed of by way of trade in any form, binding or cover other than that in which it is published, without the prior consent of the publishers.

British Library Cataloguing in Publication Data
A catalogue record for this book is available from the British Library

Printed at Brijbasi Art Press, India

Foreword

Rubbers are interesting materials and are extensively used all around us. Many Engineering marvels of today like automobiles, aerospace vehicles, machinery etc. cannot function without rubbers. Rubber have been in existence for the last 200 years and are the fore runners of the polymer industry of today. The need for technologists with knowledge of Rubbers is felt very badly in many industries.

Besides students of Rubber and Plastics Technologies, other engineers too may have to know about these materials so that materials selection can be done more effectively. Presently no good book covering the fundamentals of rubbers and their compounding ingredients is available from Indian publishers. Dr. Kothandaraman has been teaching this subject for a long time and he has attempted to write a short book which will help greatly in understanding about these vital materials.

I am glad that he has made this attempt which will help students of Polymer Science and Technology at Diploma, Bachelors' and M.Sc (Polymer Science and Technology, Materials Science, and Applied/Industrial Chemistry). This book will also help practicing technologists in Industry and will be a great asset to any Library of a University or Rubber Industry.

I wish that all individuals and organizations, dealing with Rubbers in India and third world countries benefit by using this simple book.

Dr. K. Balasubramanian
Professor, Rubber and Plastics Technology
MIT Campus, Anna University,
Chennai-600044

Preface

An index of modernity of a country can be its per capita consumption of Rubbers and Plastics. India falls way behind the advanced in this respect though the potential very much exists. The country also has the necessary technological base for this. With the economy growing fast, the usage of rubbers and plastics is bound to go up. Hence, the Indian Rubber Industry is a very fast growing one.

Advances in the fields of Rubber and Plastics technologies have changed the face of modern life. Rubbers play a vital role in automobile, aeronautics, electrical and electronics, materials handling, health care, power transmission fields besides a host of other applications. Hence, every engineer must know a little about rubber components like—oil seals, gaskets, diaphagms, grommets, belts, etc., and not the least, the pneumatic tyre. Such components must be made of the appropriate raw material otherwise break downs like oil leakage, transmission of vibrations, insulation breakdown etc. can occur.

Appropriate choice of the materials into the making of such rubber components is vital. To understand the properties, a strong understanding of the role played by polymer structure is vital and hence the book begins with a chapter on Structure-property relationships in rubbers.

The Rubber industry began with the naturally occurring material (i.e. Natural rubber - NR). This material still remains unchallenged in a few applications like heavy duty truck tyres, aero tyres though in other areas synthetic rubbers have come in a big way. Hence natural rubber will be the subject matter for the next chapter. This chapter will cover latex tapping, conversion to dry rubber, forms of natural rubber, and its modifications.

The next chapter will be on various general purpose synthetic rubbers- mainly styrene butadiene rubber (SBR), poly butadiene rubber (BR) and other such rubbers like synthetic poly isoprene (synthetic NR) and some aspects of advances made in polymerisation techniques towards development of rubbers comparable to NR like poly alkenamers.

The next chapter will be on special purpose rubbers – EPDM, butyl rubber, oil resistant rubbers - mainly NBR and poly chloroprene. The other speciality rubbers like poly acrylates, chloro sulphonated polyethylene, ethylene vinyl acetate, poly epichlorohydrin, poly sulphide rubber etc., are also covered in the next chapter.

This will be followed by the chapter on high performance rubbers mainly fluorine containing rubbers and silicone rubbers. The need for rubbers which can be processed by unusual techniques is always present. Poly urethanes, the rubbers with unusual processing techniques are also covered in this chapter.

Chapter VI will cover compound ingredients-details about curatives, accelerators, anti degradants, fillers, processing aids etc., are covered in this. The mechanism by which various curatives, reinforcements by carbon black and silica and the action of anti oxidants etc., are covered in this chapter.

The need for rubbers which can be processed like thermoplastics is being realised as the days go by - hence the next chapter will be on Thermoplastic elastomers, which is a very rapid growing area in rubber technology.

The need to understand about polymer blending-blending of rubbers with rubbers and with plastics is more acute today and a chapter on blending is added next. Principles of compounding are also covered in this last chapter.

This book will cover the important aspects of rubber materials. This will be useful for Diploma and Bachelors level students of Polymer Technology besides technologists working in Rubber Industry. This will also help aspirants of Diploma of Indian Rubber Institute exams.

Target readers: Students of Polymer Science and Technology at B.Tech, Diploma and MSc (Polymer Science/Materials Science) besides practising Technologists.

Presently there are very few books which teach the fundamentals of Rubber Materials. A few books were available from abroad-most of them are out of print now. The only available Indian book on Rubber Technology covers Processing and machinery aspects but not so much on the materials side. Thus this book will be a boon for those who want to understand Rubber materials especially from the Chemistry angle.

For understanding the subject a good understanding of the fundamentals of Organic Chemistry is essential and a good student of 10+2 can fit the bill. With this background it will be easy to understand this book.

Author
B. Kothandaraman

Contents

1

Structure–Property Relationships in Rubbers

Rubber means a polymer which is capable of reversible deformations under small applied loads. The elasticity of a rubber comes from the very nature of the chain structure of a polymeric molecule. The stress-strain curves of all rubbers are similar, though exceptions may be seen in some cases. The commonly used theories of rubber elasticity, necessary for engineering calculations, work reasonably well over a limited range of strains (roughly up to 300% elongations) but beyond this, discrepancies are seen and these are due to the chemical structural features of the polymer. Thus, the chemistry has to be considered for explaining the properties of rubbers.

The properties of a rubber are very much related to its molecular structure. Whenever a new polymer is synthesized, the first thing we may know about it, is its structure. Will it be possible to make a reasonable guess about some of its properties, from a knowledge of its structure, is the theme of this chapter.

To describe the basic structural features of polymers is beyond the scope of this book as there are a good number of books available on Polymer Science which the student is advised to refer to. For the same reason it is not proposed to cover the basics of polymerisation reactions in this book. A few basic points will be covered for refreshing the concepts already learned in polymer chemistry, to apply to Rubber Technology, at appropriate places in later chapters.

The words 'rubber' and 'elastomer' are often treated as interchangeable. Often 'rubber' means the raw polymer while elastomer, the compounded material.

If a polymer is to behave as a rubber it has to fulfil the following requirements :

1. The main chain of the polymer must be flexible (over the temperatures of use).

2. The individual chains must be capable of being crosslinked with each other.

3. The main chains must be free of weak links – otherwise they may break so easily that the rubber will be of little use to us.

1. Chain Flexibility :

This is a very important factor. How is the flexibility defined ? Is there any way by which this can be quantified ? There is a measurable property which can directly be related to flexibility.

The back bone of a polymer consists of carbon – carbon bonds (mostly single bonds). The carbon atom in a C—C bond has a tetrahedral configuration. Each single bond is capable of rotation about its axis. Since a polymer chain consists of thousands of C — C bonds, the chains can assume any configuration. This can theoretically even lead to a situation where the two ends of a chain may be the same point. With this, one can imagine to what extent each of these bonds can be straightened – this is the origin of elasticity of a polymer.

   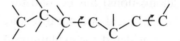   

Fig. 1.1a. Tetrahedral nature of Carbon atom   **Fig. 1.1b.** Any of these bonds can rotate through any angle leading to randam coiling   **Fig. 1.1c.** Randomly coiled polymer chain

Double bonds are by themselves rigid, but make their neighbouring single bonds more flexible. For rotation of a C — C single bond about its neighbouring bond, it must overcome a rotational energy barrier. The lowest energy barriers are associated with C — C — C bond, C — O — C bond and Si — O — Si bonds. Further, the side chains in these chains must be as light as possible. The rotational energy barriers can be overcome if the temperature in question, crosses a particular value called glass transition temperature (Tg). Thus a polymer with a low Tg value can be considered as a rubber. Tg can be defined as a temperature below which a polymer will be hard and brittle (i.e. glassy state) and above which soft and flexible (rubbery state).

The lightest side chains are H and F atoms – thus poly ethylene (PE) and poly tetra fluoro ethylene (PTFE) which contains only H and F side chains respectively, have highly flexible chains but these polymers, as we know are only plastics – this is due to other reasons which will be discussed in due course.

On the other hand, another polymer with only H as the side chain is a rubber – that is, poly butadiene (BR) especially its predominantly cis and trans forms. Poly butadiene also exists in the 1,2 structure as in medium vinyl grades, which will have a higher Tg.

The Tg of a polymer depends on the following factors :

1. *Chain stiffness*
2. *Inter chain attractions*
3. *Molecular symmetry*
4. *Copolymerisation and its types*
5. *Branching and cross linking*
6. *Presence of solvents/plasticisers*

Chain stiffness can be increased by double bonds or ring structure (as in poly phenylene or poly phenylene oxide or poly phenylene sulphide or PEEK or Poly sulphone) or ladder polymer (as is the case when poly acrylo nitrile – PAN is heated to about 250°C). Though in BR, NR (natural rubber), and CR (poly chloroprene – neoprene) a double bond is present once in 4 carbon atoms, it will not be enough to make the main chains rigid. The Tg values of the following polymers are in the following order (as the chain stiffness increases) :

Structure & Polymer	Tg
Poly ethylene	around –80°C (value is uncertain)
$-(CH_2CH_2)_n-$	
Poly butadiene	–100°C
$-(CH_2CH = CHCH_2)_n-$	
Poly carbonate (bisphenol based) 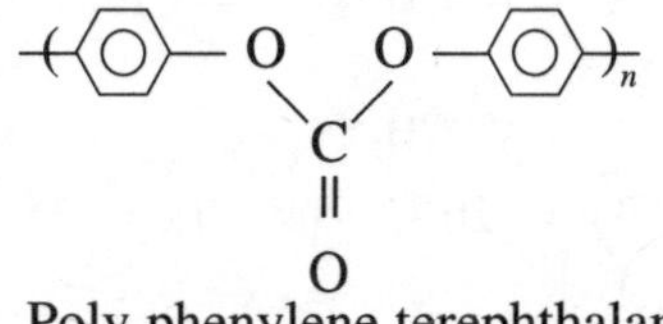	150°C

$$-\left(\bigcirc - O \quad O - \bigcirc\right)_n-$$
$$\begin{array}{c} \backslash C / \\ \| \\ O \end{array}$$

Poly phenylene terephthalamide 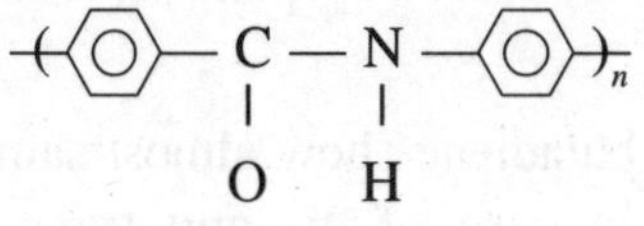	327°C

$$-\left(\bigcirc - \underset{\underset{O}{|}}{C} - \underset{\underset{H}{|}}{N} - \bigcirc\right)_n-$$

(In the last two cases, the benzene rings are present in the main chain itself – this increases the rotational energy barriers enormously).

The side chains increase the resistance to rotation of the C — C bonds. Thus when we compare poly ethylene and poly propylene, poly propylene will have a higher Tg value. This in turn will be higher in poly styrene where the side chain is the highly bulky benzene ring.

Restrictions on rotations will also be caused by *inter chain attractive forces*. For example, presence of electronegative atoms or groups as side chains increase inter chain attractions – this will increase Tg values (comparing PP and PVC, though the side chains do not differ greatly in bulkiness, PVC has a much higher Tg values (PP –20°C while PVC about +82°C). This effect will be magnified by Hydrogen bonding – as in the case of poly acrylonitrile (nearly 100°C). Conversely, increase in inter chain repulsion will lead to a decrease in Tg. If we consider the polymers of the acrylate series :

$$— (CH_2 — \underset{|}{CH})_n —$$

COOR the Tg values will show a peculiar trend as the bulkiness of R increases. If R = CH_3 (i.e. poly methyl acrylate) its Tg is 8°C while if R = ethyl (i.e. poly ethyl acrylate –24°C and poly butyl acrylate – 54°C. This is opposite the trend which we would expect—i.e. as the bulkiness of the side chains increases Tg, instead of increasing, decreases. This is due to inter chain repulsion caused by the bulkiness of the side chain. Further increase in bulkiness of R beyond the alkyl group with 8 carbon atoms again leads to increase in Tg – this may be due to crystallisation of the side chains.

The role of *moleclar symmetry* on Tg is not clear. In the case of Poly vinyl chloride and poly vinylidene dichloride Tg again decreases in the polymer containing two chlorine atoms. Similar is the case with poly propylene and poly isobutylene, though the stiffness of the chains should increase when 2 Cl or methyl sides are present, the Tg decreases.

$$\underset{Tg = 82°C}{-(CH_2-\underset{|}{\underset{Cl}{CH}})-} \quad \underset{Tg = -16\ to\ -22°C}{-(CH_2-\underset{|}{\overset{|}{\underset{Cl}{\overset{Cl}{C}}}})-} \quad and \quad \underset{Tg = -20°C}{-(CH_2-\underset{|}{\underset{CH_3}{CH}})-} \quad \underset{Tg = -75°C}{-(CH_2-\underset{|}{\overset{|}{\underset{CH_3}{\overset{CH_3}{C}}}})-}$$

This is sought to be explained by symmetry caused by, perhaps, reduction in dipole moment due to symmetry.

Similarly cis and trans forms of poly butadiene show almost same Tg (–107° and –106°C respectively) while in case of cis and trans poly chloroprenes the values are markedly different (–20°C and –45°C respectively). These are also difficult to explain.

Similarly tacticity also has, generally, little effect on Tg values.

The type of copolymerisation affects Tg values. Alternating and random copolymers show one Tg value which is in between the Tgs of the corresponding homopolymers (depending on the composition) while block and graft copolymers show 2 Tg values which are the Tg values of the pure homopolymers. SBR, a random copolymer has one Tg value (around –65°C) which can be calculated from the mole fractions of styrene and butadiene present in the copolymer (i.e. about 23.5% of styrene normally). In contrast, in SBS block copolymer which is a thermoplastic elastomer (TPE), two Tgs are seen—one at –100°C i.e. Tg of poly butadiene and another at near +100°C i.e. the Tg of poly styrene.

Cross linking reduces chain flexibility as the individual chains are chemically linked with each other. Thus as sulphur content increases in NR in a series of ebonites, Tg increases with cross link density. As we may recall, ebonite is a hard plastic. The rigidity increases with cross link density.

Branching in chains affect Tg values but the effect is not so straight forward. Branching increases free chain ends which should depress Tg while the branches themselves may act like heavy side chains – these may balance each other. However, the free chain end effect predominates often.

Plasticisers and solvents penetrate the space between the chains and this will increase flexibility of chains – thus Tg will decrease. In case of PVC plasticisation is used to reduce Tg and convert it to a rubbery polymer– Tg decrease from 82°C to, nearly room temperature.

Ability to be cross linked:

This is another important factor. Poly isobutylene has a low Tg of about –75°C but cannot be useful as a rubber as it cannot be cross linked. Hence it is copolymerised with isoprene to be useful as a rubber (butyl rubber). Cross linking may be by sulphur or peroxide or metal oxide or any other chemical or it can also be physical as in TPEs. Physical cross links serve the same purpose as chemical cross links by preventing chains from slipping past each other at ordinary temperatures.

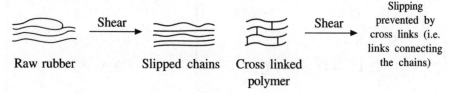

| Raw rubber | Shear → | Slipped chains | Cross linked polymer | Shear → | Slipping prevented by cross links (i.e. links connecting the chains) |

Fig. 1.2

Absence of weak links:

This is a minor requirement in the sense that any polymer with weak links in the main chain will be useless (either as a rubber or as a plastic) as it will have poor life. Examples of weak links are peroxide ($-$ O $-$ O $-$ links). This must be avoided in any polymer chain. Double bonds in the main chains are also weak links – hence rubbers with more double bonds cannot withstand higher temperatures. Similarly C $-$ C as in PVC can also be

$$\begin{array}{cc} | & | \\ H & Cl \end{array}$$

considered as a weak link as it can lose HCl easily even at fairly low temperatures which causes double bonds to form rapidly and hence rapid degradation. Hence PVC always requires stabilisers in its compounds.

Effect of structure on other properties:

Heat resistance: Heat resistance depends mainly on bond dissociation energies. Single bonds like C $-$ C or Si $-$ O $-$ Si or C $-$ O $-$ C have high bond dissociation energies, while double bonds weaken their neighbouring single bonds as they can lose H easily and form free radicals which can be easily stabilised by the double bonds besides other factors. This can lead to cross linking or degradation reactions. The details will considered in the chapter on compounding/anti oxidants. In case of silicones, the high heat resistance may be due to lack of an energetically favourable pathway to a stable degradation product (i.e. silica).

$$\text{--(CH}=\text{CH}-\text{CH}_2\text{--)}_n \xrightarrow[\text{(or heat)}]{\text{light}} \text{--(CH}=\text{CH}-\dot{\text{C}}\text{H}\text{--)}_n \leftrightarrow \text{--(CH}=\text{CH}-\dot{\text{C}}\text{H}\text{--)}_n$$

Fluorocarbon rubbers are highly heat resistant due to shielding effect of fluorine atoms attached to the main chain carbon atoms.

Benzene rings in main chains as in poly phenylene sulphide or aromatic poly amides or ladder structures impart heat resistance due to their highly rigid structures.

Table 1.2 Bond dissociation energies

	Bond	Approx. bond dissociation energy (kg mol^{-1})
1.	$-$ O $-$ O $-$	145
2.	$-$ S $-$ S $-$	270
3.	C $-$ C, C $-$ O, Si $-$ O, C $-$ C	325–375
4.	C $-$ F	425
5.	C $=$ C	600
	C $=$ O	740
	C $\equiv$ N	900

Effect of Structure on low temperature flexibility:

The factors to be considered here are Tg and crystallisation. Tg has already been dealt with, in detail earlier and hence now we will look into crystallisation. Crystallisation leads to close packing of the chains and hence restrict rotation of the main chain bonds. Thus poly ethylene, poly propylene, PTFE and trans poly isoprene are all plastics though they have low Tgs.

In polymers, crystallisation is never 100% — often the crystalline polymers may be pictured as having crystalline regions connected to each other by amorphous zones. Factors which lead to crystallinity may be light side chains, regular structures like uniformly *cis* (or *trans* etc) or tacticity (iso and syndiotactic) or alternating copolymerisation. Crystallisation does not affect Tg but increases melting point (Tm). If the Tg:Tm (in $\overset{\bullet}{K}$) ratio is around 2 : 3 such a polymer may be amorphous. Mostly rubbers fall in this category. If the ratio has a smaller value as in HDPE or PTFE or trans poly isoprene, the chains may be having a symmetrical structure. Another thing to note is that the values of Tgs of crystalline polymers are uncertain.

As far as rubbers are concerned, crystallinity is certainly an undesirable phenomenon. Without exception no rubber is a crystalline polymer, in normal conditions. This should not be confused with a property called strain-induced crystallisation seen in NR, CR and IIR (butyl rubber). These polymers have regular structure (NR has 100% *cis* structure, CR predominantly *trans* structure) and hence on stretching, the chains tend to align and then pack closely, leading to crystallisation. Of these, NR and CR also crystallise at low temperatures. IIR does not crystallise on cooling. Strain induced crystallisation confers some advantages in mechanical properties. This will be discussed in a later section.

In general a polymer is useful as a rubber at temperatures above Tg +30°C. Thus NR should be a useful rubber down to about –40°C. Crystallisation on cooling limits the lower temperature use of NR at –22°C instead of –40°C, as the polymer will become stiff (this is reversible as once the temperature is increased, this effect will vanish). For the same reason, the lower service temperature range of some grades of CR may be close to 0°C. In other rubbers, the lower service temperature limit will be about 30°C above Tg.

Crystallisation is usually caused by regular structure – this prevails in NR, CR and IIR. Most of the other rubbers are random copolymers and hence they will have irregular structure and hence will not crystallise. In case of chlorination or chlorosulphonation of poly ethylene, the resulting polymers, due to the addition of the bulky side chains, Cl and SO_2Cl groups respectively, will disrupt crystallisation and hence these polymers will become rubbery. Similarly copolymerisation of ethylene with propylene or vinyl acetate or methyl acrylate will destroy crystallisation and hence these copolymers will be rubbers. In case of vinylidene fluoride, copolymerisation will lead to rubbery polymers.

Effect of Structure on chemical reactivity: A polymer must be reactive enough to be capable of cross linking. Thus NR and BR will cure fast as there is a double bond once in every 4 carbon atoms in the main chains. In case of butyl or EPDM rubbers, the number of carbon atoms between two double bonds will be very large (a few hundreds)—hence they will cure slowly. SBR and NBR will cure slower than NR but much faster than IIR or EPDM (in these cases the number of carbon atoms between main chain double bonds will be around 8–10). In all these cases, the double bond is the cure site (for sulphur or peroxide cure).

For other rubbers, cure sites will not be double bonds but other atoms or groups – in CR, it will be the C linked to Cl in the 1, 2 or 3, 4 structures (the Cl in the 1,4 structure is not reactive for any curing reaction). Other rubbers with Cl containing C as a cure site are epichlorohydrin rubber, poly acrylate rubbers or chlorosulphonated poly ethylene (hypalon or CSM), chlorinated poly ethylene etc.

Reactivity of the rubbers also lead to modification of properties through chemical reactions. Thus many double bond reactions can be used on NR and convert it to polymers with different end use properties.

Some reactive groups may also lead to rapid degradation of polymers especially double bonds – thus NR and BR have poor heat resistance, oxidation resistance, ozone resistance and weather resistance. Double bonds can be easily attacked by oxygen in presence of heating or ozone or UV light – thus NR, SBR, NBR and BR will need protective chemicals against attack by heat, ozone and UV light – this role is fulfilled by anti oxidants, and anti ozonants. Carbon black will protect rubbers against UV radiation, present in the early morning sunlight.

Polymers with no double bonds and other reactive groups will have exceptional chemical resistance e.g. IIR, EPDM, CSM. Fluorocarbon rubbers are heat resistant due to fluorine atoms attached to C atoms. Fluorine atoms 'shield' the main chains from being attacked by chemicals.

If a polymer contains ester or amide groups in the main chains, they may be prone to hydrolysis (common examples are not found in rubber technology except for thermoplastic rubbers of the type, poly urethanes based on polyester polyols and polyesters, while in the plastics world, poly amides, poly carbonates and polyesters are common plastics prone to hydrolysis by strong bases).

$$\sim\!\!\sim CONH \sim\!\!\sim \xrightarrow[\text{H}_2\text{O}]{\text{OH}^-} \sim\!\!\sim COO^- + H_2N \sim\!\!\sim$$

(Polyamide)

$$\sim\!\!\sim COO \sim\!\!\sim \xrightarrow[\text{H}_2\text{O}]{\text{OH}^-} \sim\!\!\sim COO^- + HO \sim\!\!\sim$$

(Polyester)

In poly acrylates or EVA or EMA rubbers, the esters are in the side chains and hence these rubber will resist hydrolysis.

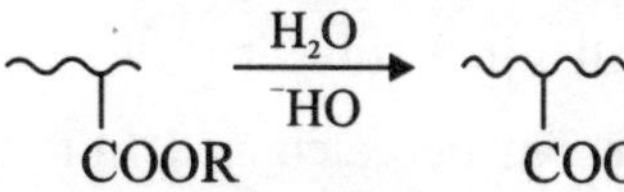

$$\sim\!\!\sim \underset{COOR}{|} \xrightarrow[\text{HO}]{\text{H}_2\text{O}} \sim\!\!\sim \underset{COOH}{|} \quad \text{(Only side chain is affected here)}$$

(as in EVA, EMA)

Benzene rings may undergo substitution reactions like nitration or sulphonation in its vacant positions without affecting the polymer – as in poly styrene or SBR. In case of poly styrene the substitution of the benzene rings by sulphonate group etc. leads to production of ion exchange resin.

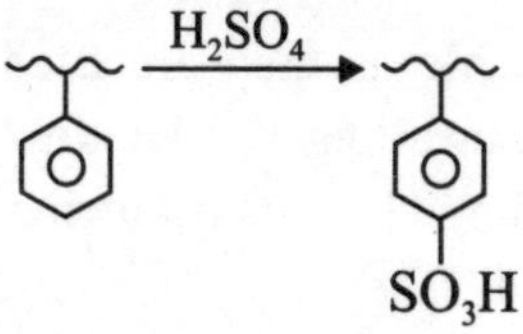

$$\xrightarrow{\text{H}_2\text{SO}_4}$$

SO$_3$H

Solvent Resistance: Solvents and plasticizers are special type of chemicals – they are essential in processing. Some rubbers are affected by some solvents. This is important when we choose polymers for hoses or seals. From our earlier education we know that polar solvents will affect polar polymers and non polar solvents will affect non polar polymer ("like dissolves like"). Polarity may be quantified using a property called solubility parameter. This property simply depends on latent heat of vapourisation and molecular weight.

Solubility parameter (M, in MJ/m^3 = $\sqrt{[(L - RT)/(M/D)]}$

where, L = Latent heat of vapourisation, R = gas constant, T = temperature, M = molecular weight, D = density.

The term under the square root is called cohesive energy density.

Non polar solvents have lower values of solubility parameter while the more polar ones, higher. More polar means more inter molecular attractive forces and hence higher latent heat of vapourisation – this increases solubility parameter.

For a solid to dissolve in a liquid, the solute molecule-solvent molecule interactive forces should be more than solute-solute or solvent-solvent molecule interactions. Thus the solubility parameters of the solvent and the polymer must be close to each other, for dissolution. When a polymer is cross linked it will not dissolve but only swell in a good solvent. Raw rubbers are difficult to dissolve in solvent because their molecular weights are large, leading to entanglements between the chains. These entanglements will prevent solubility – only when the rubbers are milled for a while they will dissolve. The problem will be severe in case of NR and CR and much less in case of other rubbers. In fact NR will not dissolve if not milled, while incase of CR for adhesive purposes, the crystalline grades may be prepared with a lower molecular weight so that they will dissolve in a solvent.

For a solvent, solubility parameter can be calculated from the latent heat values while for polymers, they cannot be vapourised without decomposition. Hence an indirect method is used to find the solubility parameter: the polymer is lightly cross linked and immersed in a series of solvents and the solvent which causes maximum swelling will be the best solvent for that rubber and its solubility parameter will be taken as the value for the polymer in question.

Swelling studies are valuable as they are used to find the cross link density of polymers as volume swelling is related to cross linking density – thus many valuable data regarding changes occurring in a rubber vulcanisate on ageing etc. can be obtained.

Solubility parameter values determine which plasticizer can be added to which polymer etc., and also which pairs of polymers can be successfully blended.

Table 1.3. Solubility Parameter MJ/m³

Polymers		Solvents		Plasticisers	
Polydimethyl siloxane	(15)	Alkanes & ethers	(15)	Paraffinic oil	(15)
EPM, IIR, NR	(16–16.5)	Esters, ketones	(16)	Aromatic oil	(16.5)
SBR	(17)	(CCl₄, toluene	(17–18)	D.O.P	(18)
ACM	18–19	Benzene, CHCl₃	(19)	DBP	(19)
Polyacrylonitrite, nylon	(29)	Phenol, formic acid	(27)		

Structure vs Electrical properties:

Polymers molecules have covalent bonds and hence are expected to be insulators. They are widely used as insulators. Modern electronics will be impossible but for this property of polymers. Many polymers are good dielectric materials – they will have a small value of permittivity because of electron polarisation – i.e. at any instance, a good number of induced dipoles are set up – i.e. nucleus of an atom attracts the outer electrons of the neighbouring molecule – thus transient dipoles (having a life of a few nano seconds) will be produced – this causes a small value of permittivity. This value will be independent of temperature and frequency of the alternating current. In polar polymers like PVC, CR, hypalon etc., in addition to the transient dipoles, a good number of permanent dipoles will also be present – this causes dipole polarisation. At low frequencies, the dipoles align with the frequency of the applied voltage. As frequency increases, these dipoles will not have enough time to align with the applied voltage frequency. Eventually a frequency will be reached where there will no time for alignment and the polarisation will be exclusively electron polarisation. For such systems, an increase in temperature will reduce viscosity of the polymer and the dipoles will move faster. This effect will be similar to the situation where a low frequency is applied. Above Tg dipoles will move faster. The incomplete alignment of the dipoles also cause loss of electric power which will generate heat (this is expressed as power factor or loss factor).

NR, SBR, IIR, EPDM are all non polar rubbers while NBR, CR, poly sulphide etc. are polar rubbers. The resistivity values of polar rubbers corresponds to a middle range of values and these are called semi conducting polymers while non polar rubbers are called insulators.

Mechanical properties of rubbers:

Some mechanical properties depend on strain induced crystallisation. Thus NR and CR will exhibit high tensile strength, tear strength etc., without reinforcing fillers (they are hence called self reinforcing rubbers). Other rubbers will need reinforcing fillers to increase their tensile strength. This does not mean that reinforcing fillers need not be added for NR—they are necessary for satisfying other requirements.

Table 1.4. Effect of filler loading on tensile properties

Rubber	Property Tensile strength (MPa)	400% Modulus (MPa)	Elongation at break (%)
NR (gum)	20	3	700
NR (filled)	30	18	600
SBR (gum)	3	1.5	800
SBR (filled)	20	1.5	600

Tensile strength was shown to be proportional to the difference between the temperature in question and Tg (i.e. T–Tg). Lower the temperature, more is the strength. Changing the test rates also affect strength values—higher the rate, more the strength. Increase in test rates and decrease in temperature both take the failure mode towards more brittle fracture.

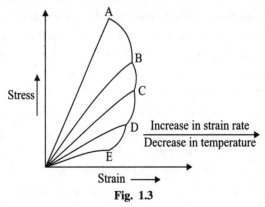

Fig. 1.3

A, B, C, D, E – Failure points for each of the curves – note that from E to A, stress-strain curves become straighter (i.e. more brittle failures)

A few more mechanical properties also depend on (T–Tg) in a similar way. **Tear strength** also depends on ability to crystallise on straining. **Crack propagation resistance on cyclic loading** also will be better if strain induced crystallisation is shown by a rubber, while crack initiation resistance depends on the ease of chain scission by stress—this means the more unsaturated a rubber, the more easily a crack will initiate. Thus, comparing NR and SBR, the two most commonly used rubbers in tyres, **crack initiation resistance** is better for SBR while crack propagation resistance is better for NR.

Abrasion resistance will be difficult to correlate with structure because the process of abrasion itself depends on a number of factors which may be contradictory to each other. Under normal conditions, the stronger the rubber the better should be its abrasion resistance, while in case of frictional heating during abrasion, a better heat resisting rubber will show better abrasion resistance. In this instance, formation of free radicals by chain scission caused by micro cutting and tearing (the main two components of abrasion), and their further propagation leading to degradation may lead to faster wear. In case of BR, its exceptional abrasion resistance is attributed to these free radicals recombining which does not seem to happen in other rubbers.

Resilience will be better when a rubber has a lower Tg, i.e. the chains are more flexible. Heat build up, an important property for tyres, will be lower when resilience is higher – poly butadiene is an exception to this principle—this will be explained in a later chapter.

Relationship between chemical structure and processing properties:

Processing properties are also important to consider before assessing the utility of a rubber. The important properties are: viscosity, extrudate characteristics, milling characteristics, cold flow and green strength and tack.

Viscosity is a very important property as it will directly determine the energy needed for mixing/mastication of a rubber. Polymer melts show non Newtonian behaviour. The more the molecular weight distribution the more non Newtonian the melt will be (i.e. shear thinning will be seen). The viscosity of an unfilled polymer will depend on the following structural factors: weight average molecular weight, chain flexibility and long chain branching. Fig. 1.4 Viscosity-molecular weight

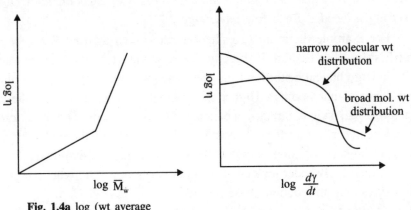

Fig. 1.4a log (wt average molecular weight)

Fig. 1.4b Viscosity-shear rate

Below a critical molecular weight, the viscosity increases linearly with molecular weight. Above the critical molecular weight, the viscosity increases rapidly. This is explained as due to more entanglements between the chains above the critical molecular weight. This explains why NR has higher values of Mooney viscosity and also why it is difficult to dissolve in a solvent.

The viscosity at a particular temperature is directly proportional to the difference between the temperature and Tg. Thus, factors which relate to Tg will also affect viscosity, $\log_{10}(\eta_T/\eta_{Tg}) = \dfrac{-17.44(T-Tg)}{51.6+(T-Tg)}$. Strain induced crystallisation will also increase viscosity. This will explain why it is difficult to injection mould NR. Injection moulding is a high shear operation. Very high shear rates will lead to sharp increase in viscosities. Long branches in a polymer on shearing will get cut from the main chain and cause lesser entanglement with other chains and hence reduce viscosity. This will explain the easy processibility of hot SBR. Higher viscosities also lead to increase in heat generation during mixing especially in presence of highly reinforcing blacks. This problem will be felt when we attempt to mix a tread compound based on NR, in a 2 roll mill.

Extrusion characteristics: One of the most important characteristics is the **die swell**. This occurs when a polymer melt comes out of an extruder, its dimensions will be larger than the die dimension (similarly calendar swell). This is due to residual stresses accumulated by the polymer as it is extruded. To be relieved of this stress the polymer melt will swell or undergo increase in temperature. Thus lesser the molecular weight (i.e. lower viscosity) or lower the extrusion temperature, the greater will be the die swell. This does not mean that we must try to minimise the viscosity of the compound. Higher viscosity is needed for a few advantages.

For example there are two phenomena known as **melt fracture and shark skin**. Melt fracture is directly due to lower viscosity – the extrudate will undergo helical distortion (this phenomenon is also referred to as elastic turbulence or bambooing). Higher the viscosity, the lesser will be the melt fracture. Shark skin is similar-appearance of transverse ridges on the extrudate or calendered sheet surface. This may due to tearing of the melt (this may occur at higher extrusion outputs). Both these make the extrudate surfaces of poor quality. Broader molecular weight distribution and higher compound viscosities help in reducing these problems.

Ability to form a band around the mill roll is a very important milling characteristic. Broader molecular weight distribution, long chain branching, blending a high molecular weight polymer with oil etc., improve this property. More polar polymer may band better around the mill roll. Some polar rubbers like CR or some grades of FKM, on the other extreme, may stick to the mill roll – they must be mixed at lower mill temperatures.

Some raw rubbers deform under their own weight even at room temperature. This is because, the room temperature for a rubber is far above its Tg. The chains can slide past each other without applying any shear. This is called **cold flow**. A polymer which exhibits more cold flow may also have poor "green" strength. This problem can be solved by using polymers with higher molecular weights (so that entanglements will occur), or by introducing a few branches/cross links. In some grades of solution SBR, a few cross links are introduced by incorporating Sn^{+4} compounds. This forms C-Sn bonds which will cross link the chains—these cross links will break on milling and thus improve banding and green strength.

Manufacture of some important rubber products need a high **tack** (ability to the compound sheet to the stick to the another sheet of the same material).This is essential for "built up" products like tyres, belts, hoses and even solid tyres etc. As it may be obvious, a rubber in which the chains can diffuse easily (i.e. low Tg) can have better tack – as in BR or NR. Once the sheets have stuck to each other, it should be difficult to separate them for producing the built up products. This condition can be satisfied only by those rubbers capable of strain induced crystallisation – thus BR will fail in the second criterion. Hence only NR and CR have good tack while most of the other rubbers will have poor tack.

Cure rates among diene rubbers will depend on the presence of double bonds. NR and BR will cure faster while IIR and EPDM will cure very slowly. Another interesting thing about rubbers is that they form ebonites when cured with high amounts of curatives. This property depends on availability of cross linking sites. This is possible only in NR, BR, SBR and NBR. Other rubbers do not have such a large number of cure sites and hence they do not form ebonites. Ebonites were an important class of insulating materials wherever rigidity was needed. Initially ebonites were superceded by thermosetting plastics due to their easier handling (e.g., phenolic resins are liquids initially – hence they have better processibility). Now a days, in some applications, even thermoplastics are replacing ebonites.

Many other properties like density, heat capacity, etc., also depend on chemical structure. They are not covered in this book though wherever necessary suitable explanations may be attempted in this book. For instance, the slightly higher density of poly chloroprene rubber may be attributed to the close packing of the polymer chains. Similar may be the case with fluoro carbon rubbers. The unusual processing behaviour of silicones may also be explained at the appropriate place.

A modern concept is to link the effect of structural variations on the properties – additions caused by the new added groups e.g. when comparing the properties of BR and NR – after all, addition of one methyl side chain in all repeating units of BR changes BR to NR (though such a conversion by a chemical reaction will be impossible) – thus we may predict the properties of NR by considering this polymer as an extension of BR having a methyl side chain through group contribution calculations. In the seventies a famous book by van Krevelan attempted to cover these aspects. Now, a few more books have appeared in this topic. Such a treatment is not attempted in this book as it may tend to wean us away from our subject.

2

Natural Rubber

The first European to see natural rubber (NR) was Columbus – during his travels in South America. He saw natives playing with balls made of dried juice of some trees – this happened in the 1490s. A French Scientist named Charles Maria de la Condamine sent the first samples of rubber to Europe and pointed out some of its possible uses, in 1736. J. Priestley named this material as Rubber due to its ability to erase pencil marks in 1770. In 1791, the first patent on this material's use to water proof textiles using rubber solutions was obtained. In 1820, T. Hancock discovered that NR could be converted to a more plastic state by mechanical shearing – this was the starting point for the two roll mill which still remains a very important processing equipment for rubbers.

In 1823, C. Macintosh patented raincoat fabric – rubber sandwiched between layers of cloth. In 1826, M. Faraday determined the empirical formula of rubber as C_5H_8. In 1839, vulcanisation was discovered accidentally by Charles Goodyear in USA. The same feat was repeated independently by T. Hancock in Britain in 1843. By the 1880s a link between the structure of NR with that of isoprene was established clearly though the exact nature of the link could not be confirmed till 1940s. In 1884, Tilden prepared a rubber like material by heating turpentine oil.

Tyres for cars were made in 1895 for a motor race. In 1899, dimethyl butadiene was polymerised to give a leather like product. In 1906, organic accelerators were discovered. In 1908, polymerisation of isoprene by sodium was observed. In 1909, reinforcing effect of carbon black was observed and found industrial application in 1912. Structure of poly isoprene was proved by ozonolysis in 1910.

In 1916 F.H. Banbury invented internal mixer for rubber. This found practical application by 1920.

After these developments many developments occurred in the field of synthetic rubbers. The concepts of "polymer" and "polymerisation" were confirmed only by the 40s though many of the technological practices for rubber materials were developed before this concept was finally accepted.

Rubbers and leathers are very important materials today and in these two fields, the naturally occurring material continues to have a dominance, while in the field of plastics, most of the natural ones have faded away giving way to synthetics completely.

Most of the rubber is used by the tyre industry (40–60%). In this aero tyres and heavy duty truck tyres are still made of natural rubber only. Thus the importance of NR cannot be overstated.

The words Rubber and elastomers can be defined by a number of ways:

Rubber – an elastic material capable of almost reversible large deformations even under small loads, while, Elastomer may be defined as a material with rubber like properties at room temperature.

Another definition says that elastomers which can be cross linked by current vulcanising agents may be called rubbers. A third definition says that rubbers are polymers, initially in a plastics state, capable of passing into elastic state by vulcanisation.

Another word for describing natural rubber was caoutchouc (meaning in the native terminology, a tree shedding tears). This word was and is continued used by Germans, Russians and Spaniards with variations.

Rubber is a polymer composed of long chains which are randomly agglomerated and entangled. These molecules can slide past each other at even moderately high temperatures. This can be prevented only by vulcanisation and hence without vulcanisation rubber will be technically not so useful. Hence vulcanisation is some times also called "curing" (curing from a disorder where the material can stiffen at around –20°C and can become sticky even at around +40°C).

Natural rubber occurs in almost 2000 plant species, out of which only a few are commercially important. The most important species is the *Hevea Brasiliensis* which is a tree which can grow up to 30 m high with a main trunk of about 50 cm diameter. It grows in tropical regions with high rainfall. Thus originally this was seen in South America in the Amazon forests and later successfully planted at South East Asia. Now most of the NR of the world comes from here. From the trunk of this tree, latex is tapped and converted to other forms, chiefly among them is the dry natural rubber.

Another species of growing importance is *Parthenium Argentatur,* a shrub of about 1m height which can be grown in less humid areas like Mexico – the NR from this source is called guayule rubber. This plant is in the form of shrubs and not trees like the other one (hevea rubber).

Latex is a white liquid like milk. It is a colloidal system i.e., suspension of rubber particles in an aqueous medium (serum). Latex may contain about 36% by weight of the rubber hydrocarbon besides proteins (about 2–3%), resinuous materials (1.6%) sugars (0.3%) and other materials. The rubber particles have a negative charge and are of spherical or pear like or oval shaped particles. The particle sizes may range from 0.1 to 4 microns. The particles are coated with a layer of proteinous and resinuous substances of hydrophilic character. When the negative charges are neutralised, the protective layer gets damaged and the rubber precipitates (coagulates). Coagulation also occurs by bacterial action. This is why latex is stabilised using ammonia solution – ammonia prevents bacterial attack. The rubber particles exhibit Brownian movement. This can be stopped by adding glycerol or gelatine or dilute salt solutions. This is used in centrifugation of latex.

The pH of latex is normally near 7–7.2. The negative charge on the particles may be explained as due to proteins. Proteins are amphoteric substances – they are affected both by acids and alkalies. They have an iso electric point of about pH 4–5. When this point is reached, the particles coalesce and readily coagulate.

Coagulation of latex can be easily done using acids or salts. Boiling, freezing out, dehydration, friction etc., can also coagulate these particles. Latex must be stabilised by increasing the pH or by the action of a lyophilic colloid.

Some substances cause delayed coagulation e.g. sodium fluoro silicate. Similar is the effect of ammonium sulphate, magnesium sulphate or diphenyl guanidine.

The exact composition of latex may vary due to the clone used. The important characteristics are:

Total solids content (TSC): rubber + non rubber solids in the latex (wt%)

Dry Rubber Content (DRC): rubber content after coagulation (wt%)

Dry rubber content may include some non rubber constituents while rubber hydrocarbon content may give the exact values of the rubber content. Density of latex may be around 0.96 g/cm³, while those of the rubber and serum are 0.914 and 1.20 g/cm³.

Hence, the density of the latex will decrease with the rubber content. Viscosity also increases with DRC. On standing the viscosity increases which can be reversed by stirring.

Latex may either be concentrated and sent for industries which make latex based products or converted to dry rubber. This book will not cover latex technology but will give more importance to dry rubber compounding and properties. Still, an attempt will be made here to give a brief outline about the various treatments to latex.

Latices may be classified by the methods used for their concentration or stabilisation.

Types of concentration methods are 1. centrifugation, 2. creaming and 3. evaporation.

Stabilisation methods are: 1. high ammonia (HA), 2. low ammonia (LA) containing, 2a. sodium pentachlorophenolate-SPP or 2b. containing zinc diethyl dithiocarbamate (ZDC).

3. Without ammonia but with fatty acid soap (FAS) for stabilisation.

We also have special latices- (*i*) physically modified – latex with high DRC and purified latex and (*ii*) chemically modified – like prevulcanised latex, freeze-thaw stabilised latex and latex grafted with methyl methacrylate.

Concentration: A few rubber products are made directly from latex – they may have better strength or ageing resistance (but with lesser thickness – hence utility is restricted only to a few products) as in condoms, examination gloves, rubber bands, catheters etc. These products may be prepared in automated production lines with high productivity and with very less energy inputs. For this if the processing unit is far away from the tree growing area, a lot of water should be removed from latex (DRC may be increased to around 60% instead of 30%) otherwise cost of transporting will become much higher. For this concentration has to be done by methods like creaming or centrifugation or evaporation or electro decantation.

Creaming: On standing for a long time, latex particles tend to separate out and form a creamy layer over the serum – this process is too slow for use – this is because the difference in densities between rubber and the serum is very low. By suppressing Brownian movement, this can be accelerated. Creaming agents are used for this purpose. They may be hydrophilic colloidal systems like a few synthetic polymers or gums or other resinuous substances which swell in water and remove and hence speed up creaming. Latex is allowed to stand for a few days and then creaming agent is added and again allowed to stand for 2–5 days. This causes separation into two layers- upper layer being the rubber rich one (60% DRC) and the lower one the serum with a little rubber.

Centrifugation: Centrifuging the latex at high speeds (8000-18000 rpm) sets up concentration gradients and this causes separation into two layers. This takes about one day. 80% of the latex products use latex concentrated by this method.

Evaporation: In the previous methods, the latex is purified from non rubber constituents. Evaporation methods concentrates without removing these constituents. Stabilisers used are alkalies. Evaporation is done in heated rotating drums, with stirring. These lattices will have smaller particle sizes and hence better for textile impregnation etc. They may absorb a little more moisture than the other lattices. They cure faster, have more mechanical stability, dries more slowly and films formed from it are more sticky.

Electro decantation: This is like electro dialysis – forcing the separation of concentrated latex and serum by the use of electricity. The advantage is that very little loss of rubber is seen in this method. Very little use is made of this type of concentration.

Stabilisation: Soon after tapping, bacterial action attacks the protein and causes coagulation of the latex. This is to be prevented as further processing of latex may take a few weeks after collection or the latex may have to be transported to far off places. The common method is to use ammonia. Ammonia can be easily removed by aeration, before the latex is used for further processing and hence highly advantageous. Ammonia content is about 0.2-0.7% by weight and is added in gaseous state with stirring. Additional stabilisers like sodium penta chloro phenolate or boric acid or ZDC. Higher ammonia additions are avoided (i.e., HA grades) perhaps due to toxicity or discolouration of the films and chemical stability. Secondary stabilisers like sodium dimethyl dithio carbamate or tetra methyl thiuram disulphide (TMTD) may also be used. None of the stabilising systems are fully satisfactory.

Physically modified latices:

(*a*) **Latex with higher DRC:** This is prepared by double consecutive centrifugation – this has higher viscosity – this is a property of advantage in dipping process as dipping can be done faster and number of dips may be decreased – thus product manufacture becomes faster. Non rubber content also decreases – this gives lighter colour and lower water absorption to the product.

(*b*) **Purified latex:** The latex is purified from non rubber constituents. This is done by multiple centrifugation and hence becomes expensive. Advantage is similar to the previous type.

(*c*) **Prevulcanised latex:** Latex is treated with sulphur and ultra fast accelerator at 70°C for 1 hr. This gives stronger films as it is cured slightly already. This process is used for dipping operations and for products where over vulcanisation can damage the film. This is not suitable for making foams. Prevulcanised latex should not be transported or stored below 0°C.

(*d*) **Latex stabilised against freezing:** At low temperatures latex may coagulate. This may be prevented by additives like sodium salicylate in combination with ammonium laurate.

(*e*) **Methyl methacrylate grafted latex (heveaplus MG):** Methyl methacrylate (MMA) may be grafted into NR in latex state and give graft copolymer with 23–50% of MMA. The commonly produced grade has about 50% of MMA grafted. These lattices can give stiffer rubber products and also with better oil resistance and better tear and ageing resistance. This may also be used as an additive to normal latex.

Advantages of Latex technology:

1. Low energy inputs for making products.

2. Only lighter equipment are needed for product manufacture.

3. Possibility of using continuous operations for product manufacture.

4. As processing does not entail mechanical action and higher temperatures, natural antioxidants may be present which may improve product life.

5. Ultra accelerators may be used without scorch danger.

6. Even if scorching (premature vulcanisation) occurs it may be harmless, as the number of cross links which may for are too few to prevent processibility.

7. No solvent is used and hence problems due to them and their recovery are avoided.

Disadvantages:

1. Processing requires stricter controls and higher worker qualifications.

2. High purity and friction free atmosphere should be observed during manufacture.

3. Transporting latex is expensive.

4. The dried waste cannot be processed as they may scorch due to ultra fast accelerators present in it.

5. Higher moisture absorption and higher modulus (this can sometime be a disadvantage).

Conversion of latex to dry rubber:

There are a number of ways of converting latex to dry rubber. For small scale production, latex may be left unstabilised and allowed to coagulate spontaneously. For industrial production, deliberate coagulation should be done.

Older methods of production:

Latex is brought to processing stations, freed from mechanical impurities, homogenised, diluted to 15–20% DRC (to allow uniform coagulation) and then coagulated using a 5% solution of formic or acetic acid with vigorous stirring. The amount of acid may be about 0.5% by weight of the rubber present in the latex. This may be done in aluminium tanks with insertable partitions. The partitions are kept after the acid is added. The coagulated rubber is passed through a series of roll mills running at even speed – this squeezes out the water present in the coagulum. The latex gradually thickens overnight to give white blocks which are washed and processed further.

Pale crepe:

The coagulated blocks are washed over several two roll mills – they are regularly grooved and rotate at different speeds. In the slit between the rolls the coagulate is torn and at the same time possible impurities are washed off with water. From the last roll mill the rubber comes out as a sheet. They are dried for 10–12 days at 30–40 C. To get a light colour, 0.5% of a reducing agent, sodium bisulphite is added before coagulation—this reduces the darkening during further processing. A grade of crepe called sole crepe is made by fractional precipitation which removes the colouring components during the first precipitation. The later precipitated fractions are taken for sole crepe.

Smoked sheets:

The majority of the NR is produced as smoked sheets as pale crepe is expensive. Here the coagulum passes through a series of roll mills under constant washing. The sheet comes out from the last mill has a thickness of about 3.2 mm. This passes through two embossed rolls which gives the characteristic ribbed appearance at the surface. The ribs improve appearance and also speeds up the drying which is done in a smoke houses for 4–6 days at 40–50°C. The smoking in modern times may be done in tunnel driers. Preservation by smoking, is still not eliminated as tunnel drying does not remove the serum completely.

Some latex, remaining in the tree dries on its own and this cannot be thrown away. They are removed before the next tapping. These (called tree lace) and some coagulated latex in the collecting cups (called cup lumps) have more impurities. These are also processed and used – they are called off grades. Additional purification operations are needed for these. They are collectively processed to give the blanket crepe. They are also used for some non critical purposes.

Modern production methods:

For countering some of the advantages offered by synthetic rubbers, modern methods have been devised. Here, the coagulation is made to give the rubber in the form of crumbs which are washed, dried and pressed to give bales and packed in poly ethylene sheets. They are easier to handle, of more exact dimensions and facilitate manipulation, transport and storage. Forming the crumb occurs either by mechanical or mechano-chemical ways. Mechanical method used rotatory knives or screw granulators – the crumbs got by this method has a particle diameter of about 5mm. Mechano-chemical method, called hevea-crumb process, involves coagulation of crumbling agents like castor oil with or without zinc stearate. The crumbling agents prevent agglomeration of the crumbs. Both methods give grades which are similar. The introduction of hevea-crumb production in plantations led to the direct preparation of oil extended natural rubber (OENR), which will not be possible with the other methods to produce NR. The advantages of OENR will be seen later.

Grading of NR: Upto 1965, grading of NR was done on purity and colour. The mechanical properties and vulcanisation characteristics of rubber was done using ACS standard recipes:

	Recipe ACS-I Parts by wt	Recipe ACS-II parts by wt.
NR	100	100
ZnO	6	6
Stearic acid	0.5	4
MBT	0.5	0.5
Sulphur	3.5	3.5

The first recipe is for grades with sufficient fatty acids while the second one is for grades containing smaller amounts of fatty acids. (Both to be cured at 141°C)

Technically classified rubber (TCR): The predominantly visual, methods for characterising NR grades was found to be unsatisfactory and hence a new system evolved. One method was technical classification – this involved use of plasticity and 600% modulus. For this, ACS-I recipe was mixed and cured at 127°C for 40 mins. For the RSS (ribbed smoked sheet) grades three groups were specified: those with red circle gave a 600% modulus of 3 MPa while those with yellow circles gave 3–5 MPa and those with blue circle, 5 MPa. This may indicate that the first grade cures slowly while the second one, medium fast and the third one, fast. The differences in cure rates were due to the variation of content of non rubber constituents.

Still later specifications were evolved one of them being Standard Malaysian Rubber (SMR). Here the first criterion was the impurities content determined analytically. On this basis, four grades were specified – colour is not specified and when light colour is to be considered, (indicated by

L grade) means low dirt content (hence, lighter colour) the number of grades become five. The next parameter is the Plasticity Retention Index (PRI) – ratio of plasticity of rubber exposed to heating in air at 140°C for 30 mins, to the original plasticity expressed as percentage. PRI indicates thermo oxidative resistance of the rubber. Grades showing values below 60% should be processed with care because they may degrade faster during service-this is because such grades higher amounts of a pro-oxidant impurity Cu^{+2}, Mn^{+2} or Fe^{+3} – these are present in the rubber due to the metabolic processes occuring in the various clones of the trees – hence they vary with the clone.

Further, Mooney viscosity values of NR increase over a period of time due to, either the phenomenon called storage hardening or crystallisation (if transported at below 10°C). The second one is reversible to heat. The first one is more serious and will be dealt with, in a later section. Thus viscosity is another important property to be specified.

Some manufacturers may like to know about the cure characteristics of the grades – for this the ACS-I recipe is used and the relaxed modulus at 100% elongation of the rubber at 140 C for 40 mins is measured – this may give a good indication of the cure characteristic of the grade. MOD term is specified - MOD 5 means 0.45-0.549 MPa, MOD 6 means 0.55-0.649 MPa and MOD 7 means 0.65-0.75 MPa. The higher value may indicate lower scorch time.

The SMR specification is completed by specifying the ash content, Nitrogen content and volatiles content. Thus we may have grades like SMR 5, SMR 10, SMR 20 and SMR 50 grades. Increase in the number may indicate increase in ash and impurities contents and reduction in PRI values. We may also have viscosity stabilised grades (CV 50, CV 55, CV 65, CV 70 etc., CV means constant viscosity and the number on the right, the range of the Mooney viscosity value.

The composition of NR is: Humidity (%), acetone extractables (%), proteins (%), ash (%) and finally 100–sum of the above give the rubber hydro carbon content (%). Humidity comes from the high humidity of the areas where the hevea tree is grown. High values may affect product appearance etc., in a very few cases though mostly it is within the limits. Acetone extractables are mainly fatty acids and other resinuous materials. Fatty acid content may affect cure rates while resins may provide natural anti oxidant activity. Nitrogen containing substances are mainly proteins – proteins may increase cure rate. Nitrogenous matter may be decreased by creaming though not completely eliminated. Ash is mainly metal ions – main dangers are posed by the heavy metal ions – copper, manganese and iron as they speed up the degradation of the rubber product during service. They may be removed by adding chelating agents while the latex is coagulated.

Properties of NR: Some of the technological properties of NR will be compared with those of BR and SBR at later sections. However some of the important points are considered here:

Chemical name is *cis* (1,4) polyisoprene (isoprene is 2 methyl 1,3

$$
\begin{array}{c}
CH_3 \\
| \\
C = CH \\
\diagup \qquad \diagdown
\end{array}
$$

butadiene) *cis* from is $-(CH_2 \qquad CH_2)_n$

Micro structure: NR can crystallise on stretching or cooling – the maximum crystallisation rate is seen at $-22°C$. This is due to its micro structure consisting entirely of the *cis* form of polyisoprene. Molecular weights are normally in the range $2.5 - 27.5 \times 10^5$ (number average) and 3 to 10 $\times 10^6$ while molecular weight distributions range from 3.6 to 11 approx., depending on the clones.

Storage hardening: The viscosity of NR increases over a period of time. This is due to the carbonyl groups from protein present in the latex which gets attached with NR during coagulation. These carbonyl groups react with each other and form temporary cross links leading to increase in viscosity. This poses a major processing problem. This can be solved by reacting the latex with a small amount of hydroxylamine hydrochloride which will neutralise the carbonyl groups and stabilising the viscosity. A number of grades of constant viscosity (CV) are available.

The protein also has other effects on NR. Vulcanisation rate, scorch, modulus and heat build up during flexing are all higher than those of synthetic poly isoprene - these are all caused by the protein present in the rubber. If the protein is removed completely these properties will be equivalent to those of synthetic poly isoprene.

Way back in 1830, it was discovered that properties changed by intensive working mechanically. Attempts to shape NR could not succeed due to its elastic recovery. Similarly additives could not be added unless the rubber was dissolved in a solvent. Further, the material became plastic on working. In the early 20th century, the link between this effect and molecular weight reduction. In 1931 it was discovered that oxygen speeded up this 'plasticisation' effect. The dependence of the extent of breakdown on temperature was found to be not so simple. The efficiency was found to reach a maximum at around 100°C.

Staudinger and Bondy in 1931 and 1932, suggested that this can be explained as involving breakdown of the polymer chains leading to reduction of molecular weights. (It must be remembered that at this point of time

Staudinger, the scientist who proposed the concept of polymers itself, could not convince the scientific world about the existence of such a material itself!)

Based on these observations, we can explain the mastication process is a mechano-chemical one:

At lower temperatures, the mechanical working leads to breaking down of chains by the chains getting cut and at those portions free radicals forming. This reduces the viscosity of the material and this causes the shear forces from the mill rolls to be unavailable for further breakdown (the rolls may slip away from the rubber). Hence the viscosity must decrease – this is seen till we reach a temperature of 80–100°C.

After this temperature is reached, oxygen plays a role – the radicals may absorb oxygen from atmosphere and form peroxy and other radicals which can lead to rapid degradation of the chains - this can cause increase in mastication efficiency with temperature.

$$(1/M - 1/M_0) / (1/M_0)$$

Without oxygen, the radicals may recombine or take part in reactions not leading to degradation. Thus the mastication efficiency will vary with temperature as follows:

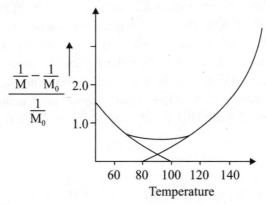

Fig. 2.1. Mastication efficiency vs temperature

The implications of such a mechanism are: (*i*) If two polymers are masticated together in absence of oxygen block copolymers must result – it has been proved to be so. (*ii*) Further, while masticating a rubber without oxygen, if a reactive monomers is added, graft copolymer must result – this is also found to be true as we can graft styrene or methyl methacrylate into NR by this method.

For speeding up the breakdown without oxygen, chemicals, mainly radical scavengers like sodium salt of penta chloro thio phenol are used in Banbury mixers – they are called peptisers.

Most of the synthetic rubbers are made to lower molecular weights-hence they may not need mastication. Further, since NR has many double bonds, it may be prone to softening by mastication. In CR, peptisation is needed in some trades.

Modified forms of NR:

Modification, here, means modification of properties by chemically reacting the polymer with other substances and changing their properties.

NR being an unsaturated polymer is expected to undergo the typical reactions of compounds containing double bonds. These reactions were at one time used to prepare new polymers with properties not normally shown by NR. With the advent of many synthetic polymers, only a few of them still have a market.

Chlorination: Olefins are easily attacked by halogens e.g., chlorine and get converted to the dichloro compounds (if there is one double bond) or tetra chloro compounds (if they have two double bonds). NR has one double bond in its repeating unit and hence a reaction of the following type is expected to take place:

$$(C_5H_8)_n + nCl_2 \longrightarrow (C_5H_8Cl_2)_n$$

This should give a chlorine content in the product as 51% but in reality the product has about 62% chlorine content. The reaction shown above, is a straight forward addition reaction which may proceed by an electrophilic addition as expected for an olefin. In the polymer, in addition some substitution reaction also occurs which may be explained as possible by an additional reaction – cyclisation which causes some substitution reaction also – this leads to liberation of some HCl which will not be possible if only addition reaction occurs.

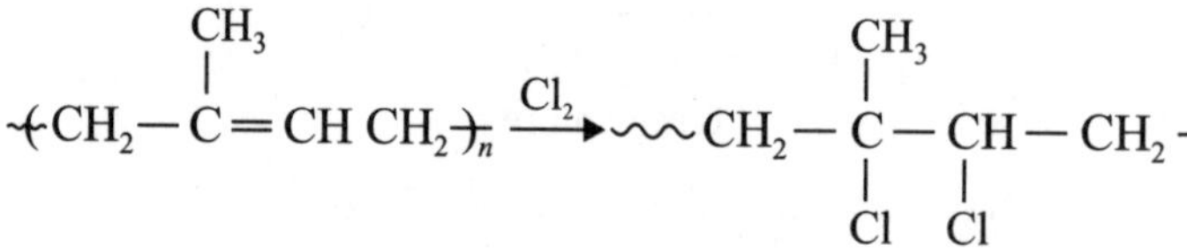

If cyclisation occurs, chlorinated NR may be

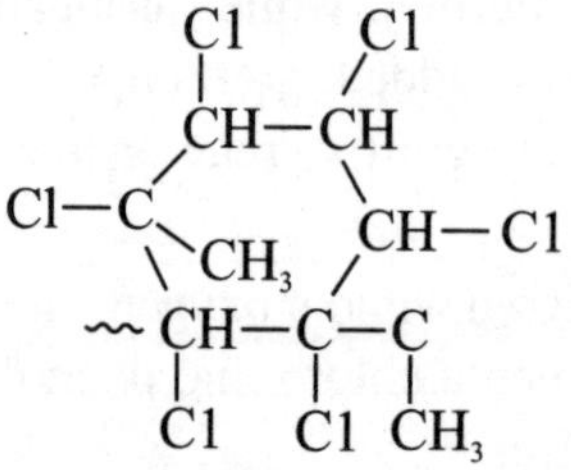

Chlorinated NR is a polar polymer and hence has better solvent resistance and since the double bond content decreases, will also have better weather and ozone resistance and since chlorine content is high it will also have flame resistance. These properties made it a very useful polymer for paints and coatings to be used in marine atmosphere (which will be highly harmful to NR). Its market may be smaller today due to the advent of CR etc, but it is still an important polymer, used as a primer for rubber-to-metal bonding.

Hydrochlorination: NR solution can be made to react with HCl gas to give an addition product, called rubber hydro chloride, which will have about 29% content. This material has been shown to be a good material for food packaging films and rubber to metal adhesives.

Hydrogenation: Like an olefin, NR can be halogenated easily. A number of catalyst systems have been proposed for this. The percentage of double bonds hydrogenated will vary with the catalyst system. Thus modification has no use these days as structurally this will be the same as an alternating copolymer of ethylene and propylene which can be made at a much lower cost by direct copolymerisation. (Similarly if polybutadiene is hydrogenated, the resulting product will be nothing but poly ethylene).

$$-(CH_2-\underset{\underset{CH_3}{|}}{C}=CH-CH_2-)_n- \xrightarrow{H_2} \ -(CH_2-\underset{\underset{CH_3}{|}}{CH}-CH_2-CH_2-)_n$$

Cyclisation: NR on reacting with an acidic substance (either protonic acid like sulphuric acid or non protonic acid like stannic chloride) can form a C^+ ion which will easily cause cyclisation reaction. Cyclisation leads to the polymer becoming stiffer, as the main chain will have ring structures, then. Thus cyclised NR will be a stiffer polymer and on blending this with NR etc, we can get stiffer compositions – these have been superceded by high styrene resins now a days.

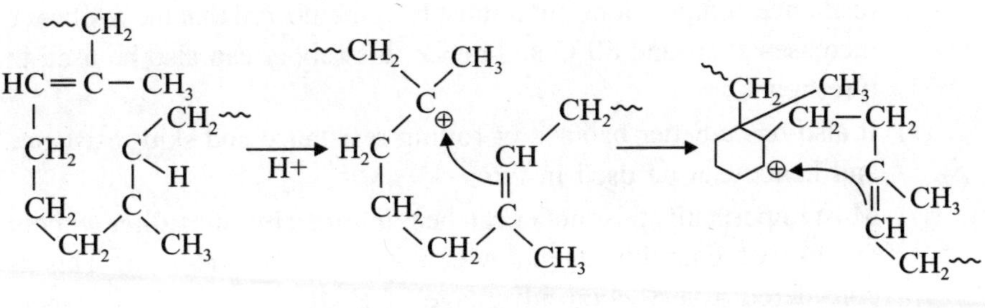

Depolymerisation: NR may be depolymerised by heating or by action of UV light. Depolymerisation can also be done in latex stage. For thermal depolymerisation of NR, the rubber may be masticated for about 40 mins and then heated on a bath, with stirring at about 250°C for nearly 4 hrs. Other methods may be to heat in presence of phenyl hydrazine etc. Photochemical degradation of NR in presence of hydrogen peroxide (by UV light for about 60 hrs) can reduce the molecular weight of the rubber to about 3000 while terminating the chain ends by OH group (this is called HTNR – hydroxyl terminated NR). Depolymerised NR is used as a reactive plasticizer for rubbers as they not only plasticize but also become part of the rubber network and hence will not cause reduction of hardness unlike the normal plasticizer. HTNR may be used as a binder for propellant systems used for rockets.

Epoxidation: This is a dramatic modification of NR, invented by Malaysian scientists as they explored ways to increase the usage of NR. Epoxidation can be a simple reaction done in latex stage:

$$-(CH_2-\underset{\underset{CH_3}{|}}{C}=CH-CH_2-)_n \xrightarrow{H_2O_2+HCOOH} -(CH_2\underset{\underset{CH_3}{|}}{\overset{O}{\overset{/\,\backslash}{C}}}-CHCH_2)_n$$

The epoxy ring formed thus confers polarity to the polymer. The extent of epoxidation can vary from 5 to 50% of the double bonds present in the rubber. The usual grades of epoxidation are 25 and 50% (grades called ENR 25 and ENR 50).

(*i*)   ENR 50 is polar enough to favourably compare with NBR of 35% acrylonitrile content – thus its solvent resistance will be as good as the normal NBR grades.

(*ii*) In addition, ENR 50 will have low air permeability like IIR and hence can be considered for tubes and tubeless tyres (the later application involves the use of a highly expensive polymer i.e., chloro butyl rubber, which can be replaced by this modified form of NR).

(*iii*) Further, since the unsaturation levels are reduced on epoxidation, the heat and ozone resistance are better than NR.

(*iv*) The epoxy ring which is the side chain is heavy and hence can reduce the resilience and hence this rubber can be used for low resilience compositions but it must be remembered that the resilience increases at around 80°C and hence this rubber can also be used in tyre treads.

(*v*)   It also has a better balance of rolling resistance and skid resistance and hence can be used in tyres.

(*vi*) Most importantly, this rubber can be reinforced by silica filler without the use of coupling agent, which is a costly chemical. This is considered a very good advantage of ENR.

For these reasons this polymer may be expected to have a good future, though it is only Malaysia which shows interest in promoting this polymer.

Heveaplus MG: This is is methyl methacrylate (MMA) grafted NR. Grafting of methyl methacrylate may be done in latex stage or by mechano chemical method on solid NR, though the, latex route is preferred mostly (either way, it is a free radical mechanism causes this grafting). The extent of MMA incorporation can vary. This modification also makes the polymer polar and stiffer, the chief use being in the field of adhesives for PVC and also for high modulus NR compositions.

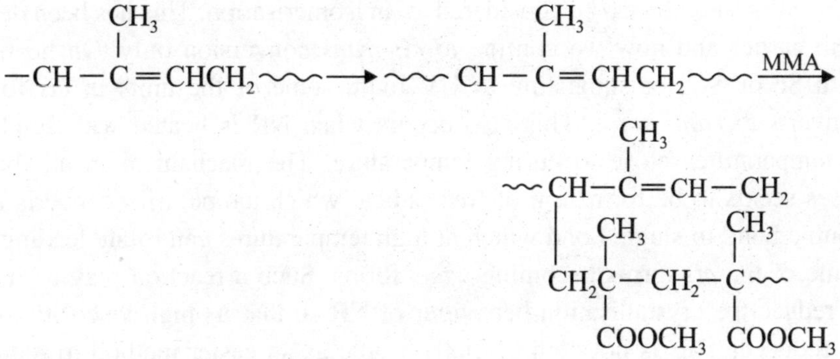

$$
\begin{array}{ccc}
\qquad CH_3 & \qquad CH_3 \\
\qquad | & \qquad | \\
-CH-C=CHCH_2 \sim\!\sim\!\sim \longrightarrow \sim\!\sim CH-C=CHCH_2 \sim\!\sim & \xrightarrow{\ MMA\ }
\end{array}
$$

$$
\begin{array}{c}
CH_3 \\
| \\
\sim\!\sim CH-C=CH-CH_2 \\
| \qquad\quad CH_3 \qquad\quad CH_3 \\
| \qquad\qquad\quad | \qquad\qquad\quad | \\
CH_2C-CH_2-C\sim\!\sim \\
| \qquad\qquad\quad | \\
COOCH_3 \quad COOCH_3
\end{array}
$$

Maleation: Maleic anhydride may be grafted on to NR either by using a free radical initiator e.g., benzoyl peroxide at 80°C or through a Diels Alder reaction (at around 220°C. Free radical route not only causes maleic anhydride to be a pendant group on the NR double bond but also some times to cross link two chains. This modification also makes the polymer polar. With this modification, NR may be blended with many polar polymers. On its own, maleated NR may show increased ageing resistance, solvent resistance besides better tear resistance.

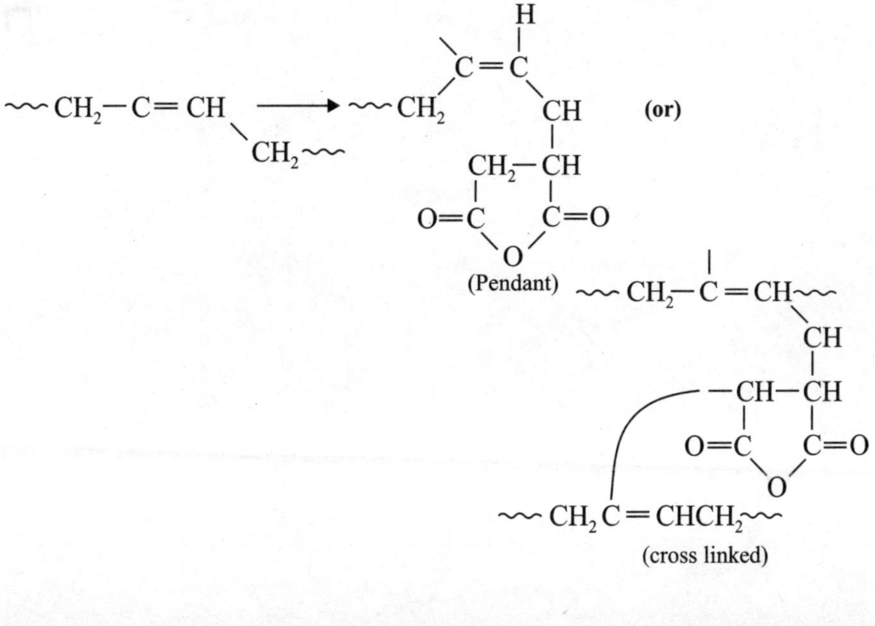

$$
\sim\!\sim CH_2-C=CH \underset{CH_2 \sim\!\sim}{\diagdown} \longrightarrow \sim\!\sim CH_2 \diagup\diagdown
$$

$$
\begin{array}{c}
\qquad\qquad H \\
\diagdown\quad | \\
C=C \\
\diagup \qquad \diagdown \\
CH \\
| \\
CH_2-CH \\
| \qquad\quad | \\
O=C \qquad C=O \\
\diagdown \quad O \diagup
\end{array}
$$

(or)

(Pendant)

$$
\begin{array}{c}
\sim\!\sim CH_2-\overset{|}{C}=CH\sim\!\sim \\
\diagdown \\
CH \\
| \\
-CH-CH \\
| \qquad\quad | \\
O=C \qquad C=O \\
\diagdown \quad O \diagup \\
\sim\!\sim CH_2 C=CHCH_2\sim\!\sim
\end{array}
$$

(cross linked)

Superior processing rubber: This is only a physical modification where NR latex is coagulated along with prevulcanised latex to give a rubber containing some cross linked structures. This gives a grade of better extrusion characteristics.

Isomerisation of NR: Isomerisation means conversion of one form to another – both forms will have the same molecular formula but for small change in arrangement of side chain or double bond etc. Considering this, cyclisation should also be considered as an isomerisation. This has been dealt with earlier and now we confine to *cis-trans* conversion only. On heating with Se or SO_2 or subjecting to UV light, some of the units in *cis* form converts to *trans* form. This also occurs when NR is heated with sulphur at temperatures close to curing temperature. The mechanism in all these cases seems to be formation of free radical which temporarily converts the double bond to single bond which at high temperatures can rotate leading to some of the *cis* forms becoming *trans* forms. Such a reaction may be used to reduce the crystallisation behaviour of NR so that its high viscosity may be reduced. This is not commercially viable as an easier method to reduce crystallisation behaviour can be by oil extension etc. In BR this reaction is of great importance, which can be dealt with in a later section.

Deproteinised NR: NR can be made free of the protein by degradation of the protein by enzmes at latex stage. This reduces the hydrophilicity in NR – this may be important in engineering products where creep, water absorption, resilience etc., may be important. Deproteinisation is important where sensitivity to protein is a matter of concern in advanced countries especially for examination gloves etc.

3

General Purpose Synthetic Rubbers

As Natural Rubber growing areas were controlled mainly by UK (and Japan during second world war), other countries started exploring possibilities of synthesizing polymers which can replace NR. In the late 19th century it was found that action of light on isoprene can give a rubber-like mass. In around 1900, dimethyl butadiene was shown to polymerise to give a rubbery material. This rubber known as methyl rubber was used in Germany for a while. Poly butadiene was also prepared but it had inferior properties compared with NR. Styrene butadiene rubber was prepared by anionic polymerization in the 30s. The real break through came in 1937 when emulsion polymerization of styrene and butadiene was discovered and commercialized. NR still is not still surpassed by a commercially viable polymer in some of its properties like low heat build up under dynamic loading, excellent building tack and high resilience.

Some ideas about polymerization:

Free Radical Polymerisation is done fairly easily. The medium of this reaction may be suspension or emulsion or solution or even bulk (i.e., no medium). This reaction does not need solvents which are expensive, can cause environmental problems besides being flammable. In emulsion and suspension polymerizations, the medium can be water itself.

These reactions can be done smoothly. In suspension polymerization, the polymer can be purified simply by filtration. Emulsion polymerization proved to be a great boon for the polymerization industry. This can lead to rapid chain growth. The main problem is, it is difficult to control the micro structure of the resulting polymer and also broad molecular weight distribution. These reduce the strength of the resulting polymer. Thus this is not a good method for homo polymers. However, free radical polymerization is highly suitable for copolymerization as highly regular structure is not expected from them. Many rubbers are random copolymers and hence easily manufactured by free radical polymerization.

Polymerisation without a medium (bulk polymerization) is rare in rubber technology – some of the plastics are made by this technique.

Ionic polymerizations are rapid and can give regular structure and narrow molecular weight distribution and hence much better mechanical properties. The polymer can be precipitated by adding non solvent or solvent evaporation. Some times removal of solvent completely, may be difficult. The biggest problem is, this method requires solvent medium. Solvents must be pure– they are expensive and cause environmental problems besides being flammable. Monomers must also be rigorously purified.

Coordination (Ziegler – Natta) polymerization yields highly stereo regular polymers – hence their utility is mainly to get BR or IR (synthetic poly isoprene). Modern polymers like poly alkenamers are also made by this polymerization.

Styrene butadiene rubber (SBR):

Styrene monomer: It is obtained from benzene which comes from petroleum sources. Benzene is reacted with ethylene in presence of $AlCl_3$ at 95°C to give ethyl benzene.

This on dehydrogenation gives styrene. The boiling point difference between styrene and ethyl benzene is not so large – thus on partial distillation etc. for separating them, there is a chance of thermal polymerization of styrene. Hence chemical methods are considered – i.e., conversion of ethyl benzene to α-phenyl ethanol, acetophenone and other products. Then acetophenone may be reduced to α-phenyl ethanol and this may be dehydrated to styrene.

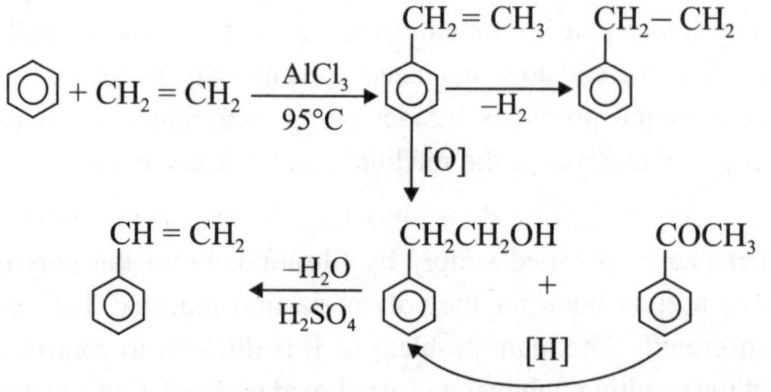

Butadiene comes from cracking of petroleum fractions containing C_4 and C_5 hydrocarbons. Butadiene can also be obtained from chemical sources e.g. ethanol which can be obtained from fermentation of sugars etc. through a series of reactions – this source is not tried now a days as it may be expensive. While styrene depends on crude oil, butadiene can be obtained from renewable sources.

Butadiene from natural sources

(1) $CH_3CHO \xrightarrow{CH_3CH_2OH} CH_2 = CH—CH = CH_2 + H_2O$

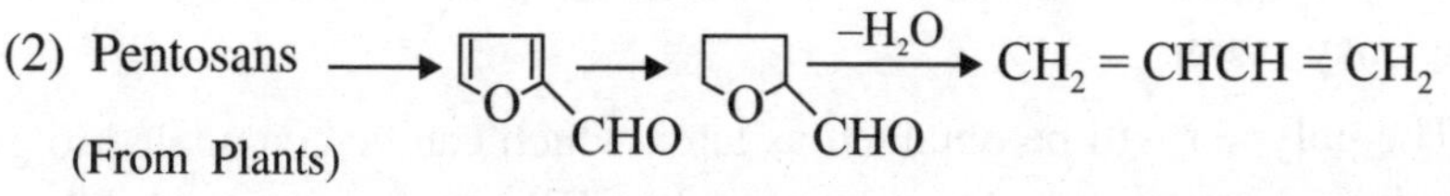

(2) Pentosans $\longrightarrow$ $\longrightarrow$ $\xrightarrow{-H_2O} CH_2 = CHCH = CH_2$

(From Plants)

This is the most used rubber in the world. Emulsion polymerization was mastered just before the start of Second World War. This polymerization was done at 50°C (this polymer is called hot SBR). In the 50s, it was found that this polymerization could be done using redox catalyst for polymerization was discovered and hence the polymerization of styrene and butadiene could be done at 5°C. This polymer had a more regular structure and hence better properties. The polymerization can be done to higher molecular weights which may be difficult to process and hence should be extended by oil (25–37 phr of oil). In the 60s, solution polymerization was discovered (anionic polymerization using alkyl lithium catalyst).

SBR manufacture: Emulsion polymerization is effected through a free radical reaction. The reaction mixture consists of the following (the table gives the recipes for hot and cold polymerizations):

Table 3.1. Polymerisation recipes for emulsion SBRs

	Component	Hot SBR	Cold SBR
1.	Styrene	75	72
2.	Butadiene	25	28
3.	Water	180	180
4.	Emulsifier system	4.5	5
5	Initiator $K_2S_2O_8$	0.3	–
	Organic peroxide	–	0.06
	Fe^{+2}	–	0.01
	Additional reducing agent	–	0.05
6.	t-dodecyl mercaptan (chain transfer agent)	0.28	0.2
7.	Inhibitor	0.05	0.05
	Conversion (%)	72	60

The initiators chosen should be soluble in water and hence the choice is persulphate or redox systems. Of these, persulphate decomposes at around 50°C. The redox systems produce the initiating radicals at 5°C itself and hence this system is called cold polymerization.

The reactions are stopped by adding short stops at conversion of 72% of the monomers in case of hot polymerization while this is at 60% in case of cold polymerizations otherwise uncontrolled side reactions like cross linking may occur.

The polymer will be obtained as latex which can be coagulated to give the rubber. Alternatively, in case of cold SBR, oil can be added before coagulation, leading to oil extended rubber. Another variation is addition of carbon black in form of dispersion in water, to the latex before coagulation leading to carbon black masterbatches. They are much easier to mix and can save energy during processing. However, they may reduce the mechanical properties as the emulsifier in the latex may reduce the carbon black – rubber interactions. Hence, dry mixing of the black is still in vogue in many places. Alternatively, the dispersion may be prepared without the emulsifier – just by stirring of the black with water and before this settles, add to the latex and then coagulated immediately.

Solution polymerization: Mainly alkyl lithium catalyst is used for this polymerization. This (anionic) polymerization is of the "living" type. This polymerization leads to a polymer of a narrow molecular weight distribution and also with very little branching. In this polymerization, butadiene may tend to polymerize much faster than styrene and hence "blockiness" may result. This may be suppressed by adding modifiers like ethers etc, or by continuous addition of monomers. Another catalyst system is alfin catalyst which consists of sodium isopropoxide (alcohol) and sodium allyl (olefin) – alcohol + olefin = alfin.

They give very high molecular weights and the polymer will also contain gels.

Table 3.2: Grades of SBR:

SBR 1000-1099	– hot, emulsion polymerized
SBR 1500-1599	– cold, emulsion polymerized
SBR 1600-1699	– cold, emulsion polymerized, carbon black masterbatch
SBR 1700-1799	– cold, emulsion polymerized, oil masterbatch
SBR 1800-1899	– cold, emulsion polymerized, more oil content than 1600 series

SBR 1900-1999	– styrene-butadiene resin masterbatch
SBR 2000-2099	– hot latex
SBR 2100-2199	– cold latex
SBR 1200-1249	– dry, solution polymerized
SBR-1250-1299	– oil extended, solution polymerized

SBR structure:

Styrene content: Generally this varies between 23 and 25%. With increasing styrene content, processibility improves and also its tensile strength, but low temperature properties and resilience decreases. Hence a styrene content of about 23–25% range is preferred. In a particular grade of solution SBR, styrene content is maintained at around 15%. This may have the advantage of better abrasion resistance and resilience and hence a possible replacement of oil extended SBR-BR blends for passenger radial tyres. This needs consideration as the prices of aromatics (hence styrene) may go up in future. Further, in very cold countries, SBR with styrene contents as low as 15% may be needed for tyres so that better low temperature resistance can be built into the rubber.

High styrene resins contain higher styrene contents (75% and above). They are thermoplastic in nature, which can soften above 50°C and flux with rubber at 85–95°C and can give higher abrasion resistance, ageing resistance and flex resistance and hence used as an additive for footwear soles, besides flooring and cable insulations. They can be called polymeric reinforcing fillers and they have lesser density than C black. There are a few more grades with styrene contents 40–50%. They are called self reinforced elastomers and are made by emulsion polymerization.

Table 3.3: Micro structure of various types of SBRs (the figures of the various butadiene isomer contents is approximate – for exact figures Ref. the standard textbooks.

SBR type	Macro structure			Micro structure			
	Branching	Gel content	Molecular wt distribution	Styrene content (%)	Butadiene cis %	trans %	1,2 %
Hot emulsion	high	high	7.5	23.5	15	45	15
Cold emulsion	moderate	low	4–6	23.5	10	55	12
Solution		nil	1.6	25			
Solution (controlled) (branched)		nil	1.5–2.0	25	25	30	22

(There is another type of solution SBR – this has more blockiness – these are easier to process and are harder but less elastic.)

(Controlled branching comes from reaction of stannic compounds with the active lithium-molecular complexes. These C–Sn bonds are weak and break on milling leading to easy processing. This also helps in preventing cold flow which is a major disadvantage of normal solution polymerization). Processibility of cold emulsion SBR can also be improved by introducing a little pre-cross linking. Hot SBR is not much in use today.

The more regular structure in solution SBR gives better mechanical properties but poorer processibility than emulsion SBR. Similarly, cold SBR has better mechanical properties than hot SBR.

The Tg of the polymer depends on styrene content and can be found by the following equation:

For hot polymerization: $Tg = (-85 + 135S)/(1 - 0.5S)$.

For cold polymerization: $Tg = (-78 + 128S)/(1 - 0.5S)$ where S is the weight fraction of the styrene in the polymer.

Such an equation is difficult to be deduced for solution polymerization.

Molecular weights of SBR can be around 80000–110000 (number average). The weight average values should be around 3–4 lakhs. For oil extended grades, $\overline{M}_n$ of about 1–1.5 lakhs and $\overline{M}_w$ about 4–5 lakhs is common – if not oil extended they will have unusually high Mooney viscosity values of about 100 units while the normal ones should be around 45–55 units. Lower values of weight average molecular weight leads to bale distortion during storage while too high values lead to difficulty in processing – as they do not break down on mastication.

Solution SBR has still not been able to replace emulsion SBR for a variety of reasons though in future some new grades of solution SBR and high vinyl BR may compete with emulsion SBR.

Polymerization of butadiene:

Poly butadiene rubber (BR) was produced by polymerization of butadiene with metallic sodium in 1930s. Its low temperature resistance was poor (it could be used down to –45°C). This grade of BR had all the isomeric forms and hence poor properties. Later on, on the lines of emulsion (co) polymerization of styrene and butadiene, homopolymerization of butadiene was also tried. These rubbers had poor processibility but may be much cheaper than solution polymerized BR. In the 40s, alfin catalyst was used but the resulting BR could not be processed easily – in those days, solutions to such problems were not forthcoming. Today, alfin catalysed BR is known but they have a small amount of other monomers too. The problem was solved with the advent of Ziegler-Natta polymerization which led to better stereo specificity. Today, both Ziegler-Natta (coordination) and anionic polymerizations are done by solution polymerization technique.

Micro structure and properties:

A wide range of catalyst combinations are available in Ziegler-Natta polymerization of butadiene. While aluminium compounds are common in all these, the transition metal component varies:

Table: 3.4. Catalyst systems for polymerizing butadiene

Catalyst system	Content of structure %			Tg C
	cis	*trans*	vinyl (1, 2)	
AlR$_3$+ Co (salt or complex)	97	2	1	
AlR$_3$ + Ni (salt or complex)	98	1	1	–110°C
AlR$_3$ + Ti (salt or complex)	92	4	4	
R$_2$AlX + Co^{+2}+ Et$_3$N	nil	95	5	
AlR$_3$ + VCl$_3$	–	99	–	
Sodium catalysed (anionic)	15	20	65	–50°C

Besides these grades, other grades like isotactic or syndiotactic 1,2 poly butadienes can also be produced by changing the catalyst systems. They are not so much useful in the conventional rubber industry. Medium vinyl poly butadienes (vinyl content 35%) are to be considered important in the future event of styrene prices going up (this polymer can be a good substitute for SBR as its Tg is much higher than that of high *cis* BR – this improves its wet skid resistance).

Alfin catalysts and cold emulsion (free radical polymerization) can give a high *trans* content in the BR. As the vinyl content increases, Tg will increase as the (side chain) vinyl, is a heavy side chain. Trans poly butadiene has a Tg of about –100°C.

High *cis* polybutadiene can crystallize easily but has a melting point of about 4°C – hence it may not have much strength. Further, at curing temperatures (in presence of Sulphur), some of these *cis* forms tend to convert to *trans* forms. This contributes to two unwanted effects: one is the low strength of the rubber and the other, the higher heat build up than expected, under dynamic conditions – hence though the resilience of this rubber is better than NR, this rubber cannot be used in truck tyre treads. Further, owing to the low melting point of the rubber, its green strength is too low that cannot even be easily mixed in a mill – the band formed around the mill roll will easily tear away – hence this rubber is not easily processible. For these reasons, this rubber is always used as blends only – with NR or SBR.

cis (1,4) polybutadiene   : $+CH_2$ $\overset{CH=CH}{\diagup\quad\diagdown}$ $CH_2\xrightarrow{}_n$

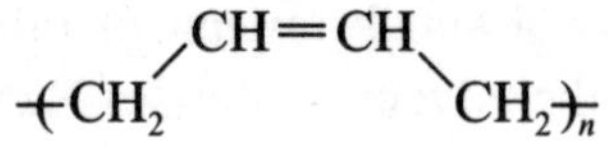

Trans (1,4) polybutadiene   : $+CH_2$ $\overset{\diagup CH=CH}{\qquad\qquad\diagdown}$ $CH_2\xrightarrow{}_n$

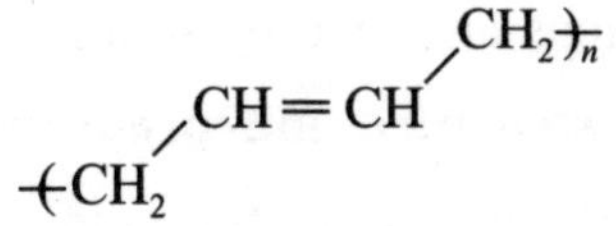

Vinyl (1,2) polybutadiene   : $+CH_2-CH\xrightarrow{}_n$
                                                  $|$
                                                  CH
                                                  $\parallel$
                                                  CH_2

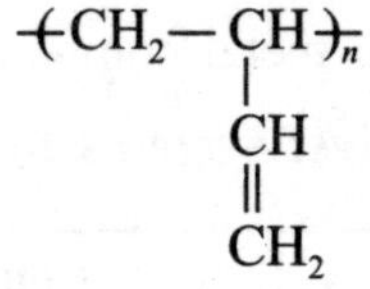

The high *trans* and high vinyl grades have higher melting points and are more crystalline – hence their utility as rubbers is poor.

High *cis* BR has about 97% and above *cis* structure while medium *cis*, about 92% *cis* – these are the common grades of BR used in the rubber industry.

Synthetic poly isoprene (IR): Like SBR and BR, isoprene can also be polymerized to give a rubber. Butyl lithium can be a catalyst – it does not leave behind, any heavy metal ion residue which can otherwise speed up the degradation of the rubber. Ziegler-Natta polymerization can also be done in hydrocarbon medium (this is a heterogenous system). $TiCl_4$ + Al Et_3 (1 : 1 ratio) can be a catalyst with addition at 4–5% by weight of the monomer. Conversion is stopped at 80% and temperature about 50°C. Butyl lithium catalysed IR has about 92% *cis* and has narrow molecular weight distribution, with no branches. Ziegler-Natta polymerized one can have about 96% or 98% *cis* content, with some branching. The rate of crystallization is lesser for IR than NR– this is due to its less regular structure compared with NR (i.e. *cis* content less than 100%).

Poly isoprenes:

cis 1,4   : $\underset{+CH_2}{\overset{CH_2}{\diagdown}}$ $C=CH$ $\overset{}{\underset{CH_2\xrightarrow{}_n}{\diagdown}}$

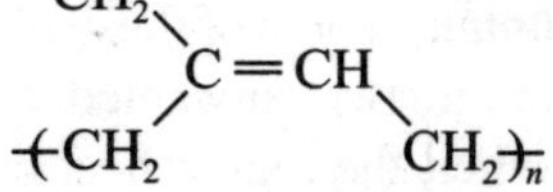

trans 1,4   : $\underset{+CH_2}{\overset{CH_3}{\diagdown}}$ $C=C$ $\underset{H}{\overset{CH_2+}{\diagup}}$

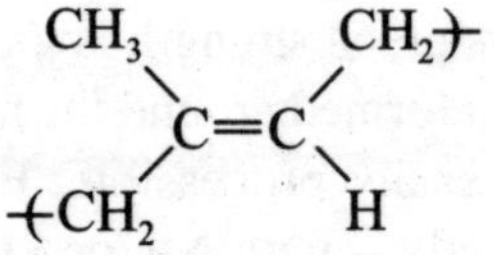

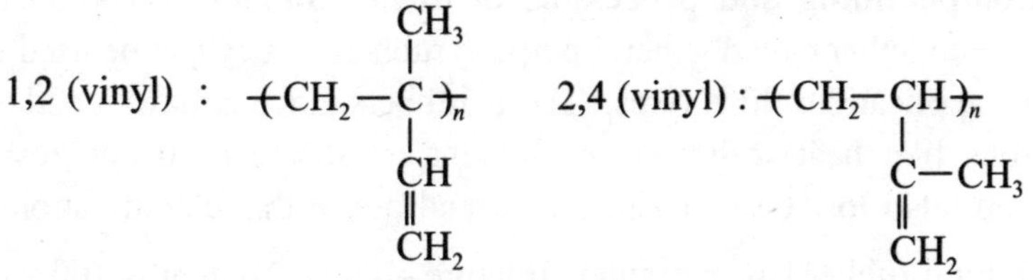

$$1,2 \text{ (vinyl)} : \ +CH_2-\underset{\underset{\underset{CH_2}{\parallel}}{\overset{\overset{CH}{|}}{C}}}{\overset{\overset{CH_3}{|}}{C}} \ )_n \qquad 2,4 \text{ (vinyl)} : \ +CH_2-\underset{\underset{\underset{CH_2}{\parallel}}{C-CH_3}}{CH} \ )_n$$

We have also seen that dimethyl butadiene can also give a rubber on polymerization. Similarly, poly piperylene (poly pentadiene) can also be a rubber but neither of these have proved to be commercially successful.

Two more polymers of poly olefin types have been synthesized and has shown to be of promise. They are *trans* poly pentenamer and *trans* poly octenamer. These are prepared through a metathesis reaction. This reaction occurs between two cyclo alkadiene units leading to a cyclic intermediate which can polymerise through a ring opening mechanism, at –50°C to 0°C in presence of Ziegler-Natta catalyst – containing (*i*) a compound of W or Mo or Ta (*ii*) R_3Al or R_2AlCl_2 (*iii*) activator like epichlorohydrin or 2-chloro ethanol.

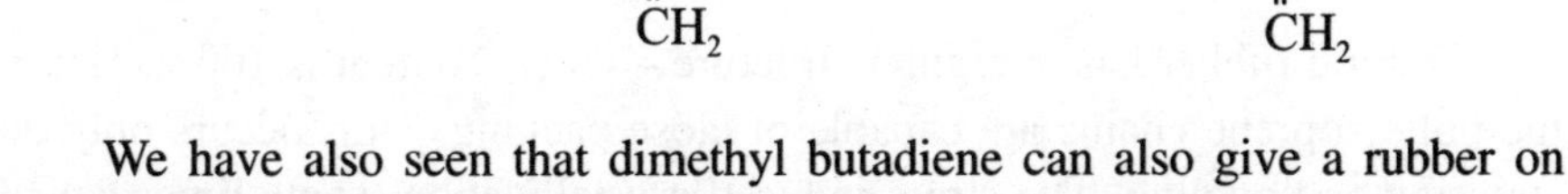

$$\rightarrow +CH_2CH_2CH_2CH_2CH = CHCH_2CH_2CH_2CH_2)_n$$

An aromatic equivalent of poly alkenamer is called poly norbornene. The monomer is prepared by Diels Alder reaction between ethylene and cyclo pentadiene. The monomer is a bicyclic ring which can polymerize by ring opening polymerization to give a poly olefin containing cyclo pentane ring structure in main chain – this makes its Tg to be 60°C.

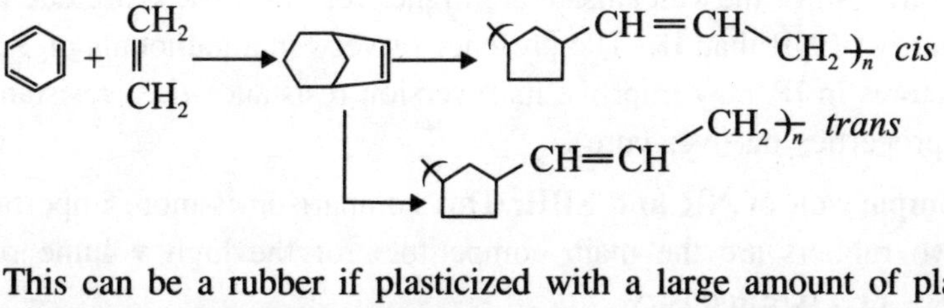

This can be a rubber if plasticized with a large amount of plasticizer (like plasticized PVC). This rubber can be cured by sulphur and accelerator system while compounding can be done in a PVC compounding machines as hardness can be controlled by plasticizer content as in plasticized PVC.

Copolymer of butadiene with vinyl pyridine can also be a rubber but its use is restricted to the area of adhesion promoter, (in combination with resorcinol formaldehyde resin) between nylon tyre cord and NR.

Compounding and processing of diene rubbers: The rubbers seen above are together called general purpose rubbers – they can be used in high volume applications like tyres. These rubbers do not have good special properties like heat resistance or solvent resistance or ozone resistance. These are also low cost raw materials and hence this classification.

Natural rubber has a regular structure – its *cis* content is 100%. Hence the poly isoprene chains are capable of close packing – this occurs only on stressing and cooling. The stress-induced crystallization (some times called strain induced crystallization) is a very important characteristic for NR which sets it apart from the other general purpose rubbers. This affects both processibility and the end use properties.

Comparison of NR and IR:

It is tempting to compare NR, SBR and BR for various properties. Before this, a comparison between NR and its synthetic alternative (IR) should also be done. The differences between NR and IR are minor but considerable in some aspects come from two factors: micro structure and protein content. As we know, NR has only the *cis* isomer in its chain structure while in IR a small amount of *trans* and 1,2 and 3,4 structures are possible.

Cure rate, tendency to scorch (premature vulcanization), modulus of the vulcanisate and heat build up during flexing are the four properties which are higher for NR than IR. The second one and the fourth one are negative points for NR. The differences between NR and IR are due to the protein mainly. The protein can work like a heat-fugitive cross link and hence higher modulus etc. Lower scorch time and cure times are due to amine groups associated with proteins.

Cold flow, green strength, festoon sagging of uncured compound and hot tear strength of the vulcanisate are higher for NR – these are due to the stereo-purity of NR than IR. The presence of very small amounts of 1,2 and 3,4 structures in IR may improve its reversion resistance (i.e., resistance to loss of properties on over cure).

Comparison of NR and SBR: This comparison is more important as these two rubbers are the main competitors for the high volume rubber business – i.e., tyre industry.

In processing behaviour, the high green strength of NR helps in better extrusion properties (this effect becomes less important as extrusion characteristics will be more driven by the amount and type of fillers in the compound). The high green strength and viscosity are due to the regular structure and the higher molecular weight of the raw rubber. Tack is better in NR than SBR because, NR has more mobile chains than SBR (lower Tg) and also strain induced crystallization ensures stronger attraction between the sheets for NR – thus it is easier to make built-up products from NR than SBR, though this property for SBR may be improved by adding tackifiers. Similarly, adhesion with metals or fibres is better for NR than SBR. It must be remembered that NR is a good candidate raw material for adhesives while SBR has much lesser utility in the adhesive industry.

Banding around the mill roll is better in NR than SBR. Energy needed for mixing and extrusion is lesser for SBR than NR – this is again due to higher green strength of NR.

Also, heat generation during mixing is higher in NR compared with SBR – again the reason may be traced to the higher green strength of NR. NR is considered to be a "self-reinforcing" rubber due to its ability to crystallize on straining while this property is lacking in SBR. In SBR, strength may be improved by adding reinforcing fillers.

Since there are lesser number of double bonds in SBR than NR (due to the styrene units in SBR), two important effects are to be seen: slower cure rate and longer scorch time for SBR compared with NR. The proteins speed up scorch and cure in NR still further. Still, blending of NR and SBR in any proportion does not pose any problem. The accelerator : sulphur ratio may be altered for these rubbers – NR may be given more sulphur and slightly less accelerator compared with SBR. Reversion resistance is much better for SBR than NR.

When comparing the end use properties, NR is better than SBR in tear resistance, tensile strength, modulus and hardness – all these are due to strain induced crystallization in NR. Resilience is higher due to the lesser Tg of NR. Similarly, heat build up during flexing is better (i.e. lower) for NR than SBR. The heat build up of SBR is so high that it is not recommended for truck and aircraft tyres. SBR-BR blends are commonly used mainly in passenger car tyres. Carcass compounds for many types of tyres will contain more NR due to its better tack. The Tg factor also makes SBR better in road grip properties (wet skid resistance) than NR while it also increases rolling resistance of SBR. Thus, as speeds of cars are likely to go up, more usage of SBR in car tyres may be foreseen.

As far as heat resistance is concerned, SBR can stand higher temperatures than NR – this is because, of lesser double bonds in SBR. For the same reason, ozone and weather resistance are also better for SBR. The same factor, further makes SBR better resistant to fatigue (crack initiation) resistance as crack initiation occurs by chains getting cut during flexing – a rubber with lesser number of double bonds can resist this better. Similarly, under some conditions like involvement of frictional heating at the abrading region BR should show better abrasion resistance than NR.

NR, SBR and BR: BR cannot be compared with NR or SBR for the simple reason that its poor mechanical strength and poor processibility makes it impossible to be used alone. It is always blended with NR or SBR.

BR has better ageing resistance than NR and SBR. Presence of BR in tread compounds improves resistance to groove cracking of the tread. The most important characteristic of BR is its outstanding abrasion resistance. Its Tg is lower than NR because of its side chain being the lighter H than the methyl in NR. Hence its abrasion resistance should be better. There is another reason for the better abrasion resistance – i.e., while abrasion occurs, free radicals are generated by the chains getting cut in the abrading surface. They recombine in BR, avoiding degradation, but not in NR or SBR. This may also improve fatigue resistance. Its resilience is higher than that of NR but this does not give the lesser heat build up as expected. This has already been explained as due to *cis-trans* conversions during cure and dynamic conditions.

This fact can be exemplified with compounds prepared under similar conditions and with same ingredients with these rubbers and testing for resilience (using Lupke Pendulum, in %) and heat build up (using Goodrich flexometer, in °C) – the values are: BR (64% and 33°C, NR 60% and 20°C, IR 58% and 18°C, and SBR 40% and 40°C.

Road grip property is poorer than SBR and NR. Its reversion resistance and ageing resistance are better than NR. The entanglements are more in BR – this leads to its capacity to take in more oil and carbon black than many other rubbers.

Poly alkenamers are better than NR in all properties. This is due to their lower Tgs (because, the side chain is only H) and strain – crystallisability. They have better heat and ozone resistance due to lesser number of double bonds than NR. Their Tg depends on their *cis* and *trans* contents - the poly pentenamer with 99% *cis* content has Tg of about –114°C while the one with 85% *trans* content has Tg about –97°C. However, the Tm for these are –41°C and +18°C – obviously the former cannot be useful as its Tm is very low. The latter grade maybe more useful, among poly octenamers grades of *trans* content 62% and 80%. Poly octenamers can be blended with SBR to improve their viscosities and strength. They may not be used alone as the molecular weights of the available grades are low and hence processibility of the polymer without other rubbers will be difficult.

Both these grades can blended in tyre formulations at 10–30% levels. Cure compatibility with SBR is good. The main draw back is that these rubbers are unlikely to come cheap in the near future due to high cost of the monomers.

Poly norbornene will give stiffer products – this may be an advantage where this property is required. It can also give better heat resistance. Further, it can accept very large loading of fillers and plasticizers without losing stiffness much. The main area where this polymer is recommended is where the rubber product manufacturer does not have conventional equipment for rubber mixing but only a PVC compounding machine. Another area may be where its damping characteristics are needed like mountings.

Uses of the general purpose rubbers:

SBR is mainly used in tyre industry. Compounds of SBR and its blends with BR find extensive uses in tyre treads for passenger cars. In carcass compounds and in some truck tyres blending of SBR with NR may be done, with good results. SBR is also used in moulded/extruded mechanical goods where the service requirements are not so stringent, footwear, cables, conveyor belts etc. NR is used extensively for production of heavy duty truck tyres and aircraft tyres.

As seen earlier, poly butadiene is almost never used alone. Its presence in blends with NR or SBR improve their abrasion resistance and groove cracking resistance. Blends containing BR may also be used in products like conveyor belts, footwear etc.

Modified forms of rubbers:

Liquid rubbers:

Liquid rubbers mean rubbers with low molecular weights of a few thousands – this is not to be confused with latex as in latex, molecular weight of the dispersed rubbers is high – a few lakhs which on coagulation gives solid rubbers.

Many rubbers of today are prepared in the form of a liquid – e.g. NR, BR, CR, NBR, poly sulphide, silicones, and polyurethanes. Among these, silicones and polyurethanes are discussed separately in their respective sections.

Liquid rubbers may be classified into two: first generation liquid rubbers (i.e., those without reactive end groups) and second generation liquid rubbers (with reactive end groups).

Among the first generation liquid rubbers, NR, BR, IIR, SBR, NBR and poly isobutylene, EPDM etc, are available in liquid forms, from a number of sources. Most of them will have molecular weights from 2000 to 40000. Viscosities may range from 50000 to a few lakhs centi poises.

They can be mixed in relatively lighter machinery and cast into the required shape and cured. They require higher cross linking agents in their compounds – their cross link density has to be more than that obtained from conventional rubbers – to achieve satisfactory level of properties. Hence, more sulphur and accelerators are needed. The liquid diene rubbers can also be cured at room temperature by dioximes + PbO_2. However, even after all these, the resulting 3 dimensional network structure is less perfect than that obtained from solid rubbers – hence the properties – tensile strength, flex cracking resistance, elongation at break etc., are poor.

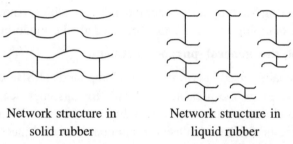

Network structure in
solid rubber

Network structure in
liquid rubber

The uses of these materials are those where the poor properties can be tolerated – as in coatings, caulks, sealants, models, toys etc. Another common use is reactive plasticizer for NR or NBR where the liquid rubber can help in processing while the polymer will also cure and become a part of the network structure – hence lesser deterioration of properties compared with a conventional plasticizer.

These rubbers may be manufactured by polymerization to give low molecular weight systems or by degrading high polymers. Example of the later is liquid NR which can be degraded thermally or photo chemically. Thermal depolymerization of NR may be done at around 240°C for a few hours. In this procedure, a few double bonds may oxidize to ketone or other carbonyl groups. Photo chemical degradation of NR may be done in presence of hydrogen peroxide to give hydroxyl terminated liquid NR.

Second generation liquid rubbers: These liquid rubbers contain reactive groups like OH, SH or COOH or amine etc. These rubbers can be cured and chain extended as in the case of urethanes. Hence they may give a better network structure and hence better properties than the first generation liquid rubbers. This is still unsatisfactory.

Terminally active liquid nitrile rubbers are very famous products from Goodrich Chemicals, USA – their main use is in toughening of epoxy resins in adhesive and composites formulations. They may be chain extended using epoxy resins and hence a few rubbery products like sealants etc., can be made.

Examples are: carboxyl terminated butadiene acrylonitrile copolymer (CTBN), amine terminated copolymer (ATBN), mercaptan terminated copolymer (MTBN) and even vinyl terminated copolymer VTBN (for toughening unsaturated polyester resins).

$$HOOC \text{-}(CH_2CH)\text{-} \text{-}(CH_2-CH=CHCH_2)\text{-}COOH \text{ (CTBN)} + \triangle_O$$

$$\begin{array}{c} | \\ CN \end{array} \longrightarrow HOO \; (\sim COO - CH_2 - CH \sim$$

$$\begin{array}{c} | \\ OH \quad \text{chain extention} \end{array}$$

Hydroxyl/carboxyl terminated poly butadiene (HTPB/CTPB) is also made by a few companies – they can be cured by isocyanates and the main use is for liquid rocket propellant binders. Similarly HTNR can also be used. HTPB and HTNR can be cured using di-isocyanate and chain extender containing aromatic groups – they give self reinforcing rubbers and hence good mechanical properties maybe obtained – these rubbers must be considered alongside poly urethanes (PUs). Polyurethanes can be processed in liquid state and still give good mechanical properties – it is possible to make tyres for lighter vehicles and also for earth movers etc. by PUs and its attendant unique processing techniques – they have not yet penetrated the tyre market due to two reasons – one the slightly poorer elastic properties and the other being the very high cost of the raw materials.

HTPB : $HO\text{-}(CH_2CH=CHCH_2)_n OH$

Regarding the other liquid rubbers, their compression set properties are poorer and this is a major dampener in their acceptability by users of elastomeric products.

Liquid rubbers can be easily processed in liquid processing equipment like RIM, which are much lighter and can be automated and are much less labour intensive – thus it is expected that compared with a typical tyre factory which requires production capacity of at least a few thousand units per day for attaining economies of scale, liquid processing plants can attain this with nearly a tenth of this capacity – thus it is tempting to think that liquid polymers will have a bright future.

However, the poorer mechanical properties and difficulties in using the typical sigma mixers etc. for compounding are major dampeners. Mixing a carbon black with a liquid rubber renders the material becoming a paste which is not easily processible unless high shear forces are given to it. This calls for low productivity machines like three roll mill for such materials – clearly the development of a third generation liquid rubbers is required for solving these two problems – till this happens, liquid rubber processing cannot become a major technology.

Powdered rubbers:

Rubbers on cryo grinding becomes a powder. At temperatures below Tg of a rubber it becomes glassy and in this condition this can be ground to a powder. At room temperature, these powder particles can recombine and the material can become a block which is the original state of the raw polymer.

This problem can be solved by cryo grinding the rubber in presence of inert fillers like calcium carbonate, mica etc, at a loading of about 10–15 phr. These are called partitioning agents and they help in maintaining the stability of the powder, without deteriorating the properties of the vulcanisate.

A powdered rubber can have many advantages –

(*i*) bale cutting operation can be eliminated

(*ii*) low energy inputs during mixing – because, it is easier to mix fillers and other chemicals when the raw material is in the powdered form – this enables faster mixing of reinforcing fillers with lesser heat generation during mixing

(*iii*) lesser wear and tear of mixing equipment and hence lesser maintenance of the machinery

(*iv*) due to lesser heat generation, more loading of faster accelerators can be done with much lesser scorch danger

(*v*) lesser heat history developed in the compound

(*vi*) possibility of using continuous mixers for making rubber compounds – this is much more difficult with the rubber in bale form.

(*vii*) degree of dispersion is much better than mixing with bale form of rubber (Points (*v*) and (*vi*) make it possible to directly feed compounds to further processing or to do processes like mixing and extrusion in the same machine at the same time).

All these, make powdered rubbers an attractive proposition – however, cryo grinding is a very expensive process as it requires liquid nitrogen. The other problem is the economics behind transportation of powdered rubbers *vis á vis* bales. Recently, the production of powdered rubbers has changed over to the more economic methods like spray drying etc., of latex. Hence powdered rubber production will see more customers in future.

Thermoplastic elastomers and reclaimed rubbers too need coverage as they are also important raw materials. Thermoplastic rubbers have a different chemistry behind their manufacture and processes while reclaimed rubber is a part of recycling of polymers – hence they will be considered in separate chapters.

⌐

4

Special Purpose Synthetic Rubbers

The high volume of rubbers produced are mostly used by the tyre industry (about 55% of the rubbers). The rest of the rubber industry consists of many other products like belting, hoses, cables, diaphragms, grommets, oil seals, engine mountings etc. Many of these products will require properties which cannot be met by the general purpose rubbers. Hence the need for special purpose rubbers has been arising for a long time and these needs have been met by the rubber manufacturers successfully.

The properties required may be low air permeability, superior heat, weather and ozone resistances, solvent/oil resistance, heat resistance with solvent resistance, resistance to powerful oxidising chemicals, flame etc. These in turn, require special purpose rubbers. Among the special purpose rubbers, three of them do not possess oil/solvent resistance. Further, one among these will be considered as a high performance rubber (silicone). The rest have solvent resistance to at least some degree. Hence, first, we consider the two non-oil resistant special purpose rubbers – namely, butyl (IIR) and EPDM.

Butyl Rubber (Isobutylene-isoprene rubber-IIR):

Isobutylene was the first alkene to be polymerized and poly isobutylene, the first polymer to be commercially developed. Isobutylene can be easily polymerized by cationic polymerization but the polymer cannot be useful as a rubber though its Tg is low. This is because this polymer has no reactive groups or double bonds which can be cure sites. The only cure system possible for such a polymer, i.e., peroxide, fails in this case as the free radical formed on the poly isobutylene chains will tend to undergo chain scission much more than cross linking (due to the steric effect caused by the two methyl side chains). In the early 40s, it was found that a small amount of comonomer, isoprene if incorporated into this polymer can allow this polymer to be cured by sulphur and accelerator. The production of this rubber began in the 40s.

Polymerization: This is perhaps, the only highly successful example for cationic polymerization, commercially. The monomer, isobutylene has a carbon atom which has two methyl side chains and hence can easily form a stable, a tertiary carbocation – thus it is easily susceptible to cationic polymerization.

$$-CH_2-\underset{\underset{CH_3}{|}}{\overset{\overset{CH_3}{|}}{C^+}}$$

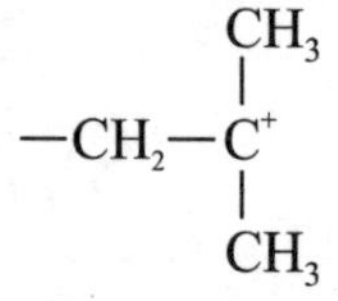

Isobutylene with a small amount of isoprene is dissolved in methyl chloride and treated with boron tri fluoride with a trace of moisture. The reaction requires a very low temperature (about –80°C), for removing the exothermal heat, and this is achieved by conducting the reaction in liquid ethylene. The polymerization is ionic and hence very fast-it takes about 2 seconds to achieve a high molecular weight.

Structure and properties:

The isobutylene units are attached head to tail always. The isoprene content may vary from 0.6 to 3 mole %. Higher mole % causes easier degradation of the polymer but speeds up the cure. The viscosities are fairly low and hence they are easy to process. Molecular weight distribution is about 3.

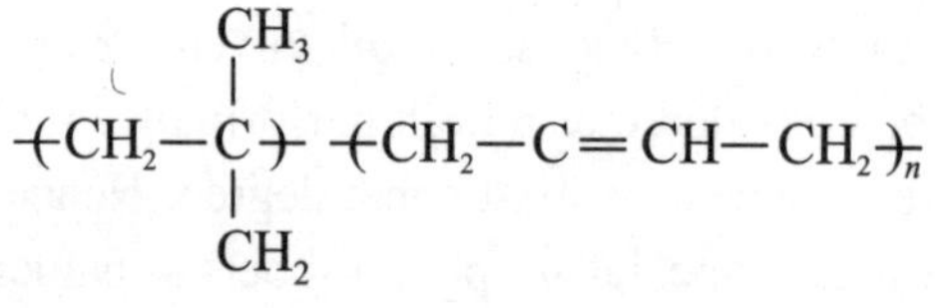

$$\underset{\underset{CH_2}{|}}{\overset{\overset{CH_3}{|}}{(CH_2-C)}} + (CH_2-C=CH-CH_2)_n$$

Butyl rubber, due to the main monomer being isobutylene (about 97%), has a symmetrical structure and hence regular arrangement. When a stretched film is viewed under X-ray diffraction studies show a crystalline pattern. The chain is a planar zig-zag in shape with the methyl side chains arranged as a helix which repeats once in 8 carbon atoms–thus it will appear as if the methyl side chains are "shielding" and over crowding around the main chain. This arrangement accounts for two of the important properties of IIR-one is the low air permeability – hence its unique market in the field of automotive tyre tubes. This overcrowding also reduces the chain flexibility – though the Tg of this rubber is around –70°C (a reflection of flexible chains), the segmental motion of these chains are sluggish – this manifests as low resilience among all rubbers. Thus for low resilience applications (damping), this rubber is useful.

$$\begin{array}{ccc} CH_3 & CH_3 & CH_3 \\ \diagdown C \diagup & \diagdown C \diagup & \diagdown C \diagup \\ \sim C \diagup \, | \, \diagdown C \diagup \, | \, \diagdown C \diagup \, | \, \diagdown \\ CH_3 & CH_3 & CH_3 \end{array}$$

The other important properties of this rubber are the high heat, oxidation, ozone and weather resistances. These are due to the very low unsaturation levels in this polymer. This factor also confers a very good acid resistance for this polymer.

Due to the reasons discussed earlier, the rubber cannot be cured by peroxides but only by sulphur and other curing systems used for diene rubbers–like oxime and phenolic resin systems. A peroxide curable grade of IIR has been prepared by introducing a few units of di vinyl benzene whose pendant vinyl group can easily cross link with peroxide.

The low unsaturation can have other implications – if contamination (even by accident) occurs in a butyl compound by NR or SBR, the curatives will act only on the NR or SBR (as IIR cures very very slowly while NR or SBR will cure much faster) and hence cure compatibility of IIR with the cheaper NR or SBR, in spite of the similar polarity of these rubbers, is not possible. To overcome this problem, halogenation of IIR is done. Mostly only chlorine and bromine are added. This reaction may be done on the rubber solution and passing Cl or Br. Bromination occurs either by substitution (free radical reaction) or addition mechanism (through C^+ ion) – both these mechanisms act only on the isoprene units. Substitution is predominant. Chlorination is also mostly substitution but at $-20°C$ or below, addition is observed.

Cure compatibility is achieved by the new cure site (i.e., halogen) added. The ZnO, which is a regular rubber additive to NR or SBR can cause cross linking through the halogen atom – this reaction proceeds independently of the sulphur cure of the NR or SBR present in the blend – thus the interference by the more reactive rubber will be avoided. Halogenation also increases heat resistance. Further, a few more curatives can be used on this rubber because of halogenation. Halogen content in these rubbers is low often 1–2.5%.

$$\sim\!(CH_2\!-\!\overset{|}{\underset{|}{C}})\!+\!(CH_2\!-\!\overset{|}{C}\!=\!CH\!-\!CH_2)_n \xrightarrow{\ Br\ } CH_2\!-\!\overset{|}{\underset{Br_2}{C}}\!=\!CHCH\!\sim$$

$$\sim CH_2\!-\!\overset{|}{C}\!=\!CH\!-\!CH\!\sim \atop | \atop Br$$

Curing of butyls:

Due to the presence of isoprene, this rubber can be cured by sulphur with accelerator. The systems are similar to those used for the general purpose, diene rubbers but since the unsaturation levels are very low, curing will be very slow. For these reasons accelerators may be added in higher dosages while sulphur, in very low amounts. Ultra accelerators can be added without any scorch danger. A typical cure system can be: sulphur 0.5 phr + TMTD/ZDC 1–3 phr, with a thiazole at 0.5–1.0 phr level. Tendency for reversion may be seen in sulphur cure systems.

As seen earlier, peroxide cure is ruled out. Other double bond curing systems like oxime and phenolic resin cures are regularly done in products requiring high heat resistance. For example, p-quinone dioxime (2 phr) with MBTS (4 phr) (to oxidize to p-dinitroso compound, which is the real curative here) can be a cure system, but this gives poor strength and resilience. MBTS may be replaced by PbO_2 for oxidizing the oxime. This speeds up the cure – this can be used to cure the liquid butyl rubber. A better curing system is the reactive phenolic resin system. Phenolic resin containing active Br atoms is used. This also gives high heat resistance – this is exploited for making the tyre curing bladder which has to constantly withstand temperatures as high as 170°C for about 20 mins for around 500 cycles. If the brominated resin is not available, ordinary resin can be used – but the formulation should contain about 5 phr of CR or any other halogen containing polymer or $SnCl_2$.

Chlorobutyl rubber (CIIR) can be cured by ZnO – curing occurs by removing the chlorine atom. TMTD can speed up the cure but also increases scorch danger – this may be replaced by MBTS. ZnO + ethylene thio urea can also be used a curative. Bromobutyl (BIIR) has even greater cure versatility and cures faster than CIIR.

Terpolymerisation of isobutylene & isoprene with a small amount of divinyl benzene gives a branched polymer, with a few cross links. This gives good green strength resistance to cold flow. Further the pendant vinyl groups help in peroxide curve, which in turn improves heat resistance eqivalent to that of henolic resin cure. This polymer is useful for sealing, cushioning, pressure sensitive adhesives etc.

Properties and uses:

Butyl rubber does not breakdown easily in milling. Thus, in presence of a promoter like p-dinitroso benzene, this rubber interacts with carbon black in a Banbury mixer at about 180°C and gets better reinforcement (this is called heat treatment of IIR). Modulus, Tensile strength etc., are increased without loss of resilience (such a phenomenon is not observed in NR or SBR).

IIR crystallizes on stressing but not on cooling. The more important properties are its low air permeability, low resilience, high heat, ozone and weather resistances and chemical (acid) resistance. This rubber is used for tyre tubes – this is due to its low air permeability. Its low resilience is exploited for vibration dampers. The resilience becomes as high as that of SBR at above 50°C. Heat resistance, ozone resistance and weather resistance are also outstanding but these properties are also provided by EPDM which is the more preferred rubber for such applications. CIIR and BIIR are similar to IIR but more cure-compatible with NR or SBR and hence it is preferred for innermost liners for tubeless tyres.The halogenated IIR can also be cured by non toxic cure systems like radiation – this may be exploited in medical goods industry. Other uses may be for conveyor belts carrying hot materials, adhesives, side walls of tyres, acid resistant tank liners, steam hoses.

EPM and EPDM:

Poly ethylene has flexible individual chains and hence can be a rubber but its high crystallinity prevents the flexibility of the polymer as a whole. It has no unsaturation and hence outstanding electrical properties, heat, ozone and weather resistance. If its crystallinity can be disrupted it can behave as a rubber. The best way to do that is to copolymerize. Thus, copolymers of ethylene with polar polymers like methyl acrylate and vinyl acetate give good rubbers–but these are polar polymers and they lose the good properties which poly ethylene can provide. Hence, copolymerization with propylene can give a good rubber, but till the 70s, no method to copolymerize these monomers could be discovered.

Attempts to copolymerize them failed – ths reason was ethylene always polymerized much faster than propylene and hence the copolymer never formed. Even the initial Ziegler-Natta catalysts failed. Finally, in the 70s, a special-Ziegler Natta catalyst succeeded in copolymerizing these monomers – it was based on $Et_2AlCl + VOCl_3$.

The medium for this reaction is a hydrocarbon or halogenated hydrocarbon solvent. In a particular company, propylene in liquid state itself is used as the solvent. The copolymer composition is independent of polymerization time, catalyst concentration and Al:V ratio. The molecular weight depends on these factors – increases with reaction time and ethylene:propylene ratio, and decreases with temperature, catalyst concentration.

Structure-property relationships:

$$EPM \quad : \quad +CH_2CH_2+ +CH_2-\overset{\displaystyle CH_3}{\underset{|}{CH}}+$$

$$EPDM \quad : \quad +CH_2CH_2+ +CH_2-\overset{\displaystyle CH_3}{\underset{|}{CH}}+ +diene+$$

Dienes :

CH_2=CH
|
CH_2CHCH=$CHCH_3$ dicyclo pentadiene   ENB
                                 DCPD   ethylidene norbornene
(1,4 HD) 1,4 hexadiene

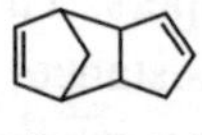

EPM (M means polymethylene–the repeating unit) has no unsaturation and hence, can be cured only by peroxide. To rectify this problem, a double bond containing monomer should be copolymerized with ethylene and propylene. The choice of the third monomer was not easy – initially, conjugated dienes were tried but they polymerized much faster than the main monomers and may form cross links (gels) – this is undesirable. Later, a few non conjugated dienes were tried – they have comparable polymerization rates and will not prevent the reaction from proceeding to high extents. Such conditions could be fulfilled only by a few systems – they have isolated double bonds and the second double bond is present either in a side chain or in a pendant ring. Thus the common third monomers in EPDM are: 1,4 hexadiene (1,4 HD), ethylidene norbornene (ENB) and dicyclo pentadiene (DCPD). Each has its own advantages and disadvantages:

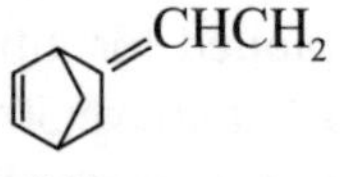

ENB: This monomer is 10 times more expensive then DCPD but is the most popular monomer – it cures faster, is easy to incorporate and is favourable to co-curing with SBR for blending.

DCPD: Least expensive but has methine hydrogen – this may cause more chain scissions – may affect cure efficiency.

1,4 HD: Provides the best ozone and heat resistance and ease for recycling.

The diene content in the rubber may be 4–5%. However, if it is necessary to blend this rubber with the faster curing diene rubbers, or to cure faster, then grades containing 8–10% of dienes may be used.

The ethylene content in the rubber is also important. The ethylene content may vary from 45% to 75%. More ethylene content leads to better extrusion behaviour – lower die swell, and better shape retention but impairs the ease of mixing. More propylene leads to easier mixing and moulding behaviour.

Ethylene content contributes to viscosity – grades of higher viscosities have better mechanical properties. Higher ethylene grades may show some crystalline regions which is reflected in better mechanical properties. The polymer chains have branches. The presence of diene in the reaction mixture leads to more branching – it can be expected that EPM may be less branched.

Another important feature in these rubbers is the location of the double bonds – they are located in the side chains unlike NR or SBR or IIR. Thus if chain scission occurs (which will be near the double bonds) they will affect only the side chains. Hence, the main chain will not be affected – thus the heat resistance should be very high – higher than IIR where the double bond is in the main chain. It is observed to be so.

EPDM has the lowest density compared with other rubbers. This is another advantage – as raw rubbers are bought in weight and rubber products, sold in volume. The Tg values are close to that of SBR and the resilience is better than that of SBR and much better than IIR.

EPDM can be extended with a large amount of oils and fillers. Mostly naphthenic oils are used as they will not impair the ageing resistance. Paraffinic oils also can work for EPDM. A large amount of carbon black can also be loaded. Thus the cost of EPDM compounds can be reduced even though the base polymer itself may be expensive. This factor, coupled with the high performance properties like heat, ozone and weather resistance make EPDM a rubber of high performance : cost ratio.

Due to its low unsaturation levels and lack of polar groups, its electrical properties are excellent.

The rubber tends to lose viscosity rapidly beyond a certain temperature – this is called thermoplastic nature – such rubbers may pose problems in mixing – in internal mixers, for this reason, an abnormal procedure called "upside down mixing" is followed – here, the carbon black and oil are fed first followed by the rubber – this helps the stock to retain some stiffness so as to enable to get the shearing action of the banbury rotors.

Properties:

The main properties which are exploited by the industry are: outstanding resistance to heat, ozone and weather, good resistance to polar substances, good resistance to steam, excellent electrical properties, good mechanical properties and abrasion resistance and flex cracking.

Peroxide curing gives the best heat resistance. Sulphur cure can also be done. Anti oxidants have better effect on peroxide cured compounds but not much on sulphur cured systems. EPDM can work as a physical anti ozonant for SBR or NR. This application does not need co-curing – the EPDM forms an immiscible phase (dispersed as small particles) which stop the ozone attack at its surface.

Applications:

EPDM is a relatively cheap polymer for the special properties given by it. At one time it was expected that EPDM may become a general purpose rubber. However, the termonomer and the cure systems are expensive and hence the rubber cannot be very cheap – however it can be made to work as well as a general purpose rubber for tyre applications. Though EPDM is never accepted by the tyre industry except for side wall applications, its uses are increasing in other industries.

Its low unsaturation level (hence low scorch tendency) and low viscosities at high temperatures make it an ideal elastomer for injection moulding. Many automotive components can be made by this process profitably. Further, it can be loaded with large amounts of fillers and hence can be extruded smoothly. Many extruded products for automobile industry are now made of this rubber – window seals, radiator hoses, welding hoses and many other profiles. Automotive tubes are made of blends containing EPDM. Tyre curing bladders too may contain some EPDM in their compounds. Liners for protection against acids, alkalies, synthetic hydraulic fluids, animal fats and polar solvents. It can withstand temperatures of about 100°C while in contact with oxygen and 160°C in oxygen free medium e.g. steam.

Another important property of this rubber is its ability to retain light colours. This property along with its high flex resistance, is exploited in its use as white side wall of a radial tyre. Its performance in weathering sheets is also high due to the same property, in addition to its low unsaturation levels (hence high weather resistance).

Whether EPDM should be called a general purpose rubber or a special purpose rubber is a matter of debate. It is tempting to classify this under general purpose rubber due to its high performance to cost ratio relative to the other general purpose rubbers. Considering its high cost it may be called a special purpose rubber. Perhaps it can be called a (relatively) low cost rubber with some special properties.

Nitrile Rubber (NBR):

It is a copolymer of acrylonitrile (N-stands for nitrile) and butadiene. It is the most important oil resistant rubber. Like buna-S, buna-N rubber was also prepared earlier but like SBR of today, the nitrile rubber made today is by emulsion polymerization. Like SBR, here too, hot and cold NBRs are manufactured. Today, mostly the cold NBR is manufactured (at 5°C) with redox initiator system. During drying stabilizers are added which work even during and after cure. If the stabilizer is not added, even at 60°C, the polymer may crosslink and form gels. The polymerization is stopped at around 90% monomer conversion.

$$+CH_2CH+ +CH_2CH = CHCH_2+$$
$$|$$
$$CN$$

Unlike SBR production, the composition of the feed and that of the copolymer may not match. Branching is less in NBR than in SBR for the same polymerization temperature. Though gel formation is possible, modern grades may not contain gels.

NBR does not crystallize on stretching or on cooling. Hence, reinforcing fillers are a must for achieving good strength. Density of the polymer increases with ACN content.

A number of grades of NBR are available – the grades differ by Mooney viscometer, acrylonitrile content, polymerization temperature and type of stabilizer (staining or non staining). Similar to SBR plasticizer-extended grades can also be prepared.

Effect of acrylonitrile (ACN) content:

Based on ACN content NBR can be classified into the following:

Type	ACN content
Low	upto 25%
Medium	30-35%
High	40%
Very high	45%

As the ACN content increases, polarity increases, compatibility with polar polymers improves, tensile properties increases, heat resistance, solvent resistance and processibility increases, while at low ACN contents, low temperature properties improve.

Mooney Viscosities vary from 35 to about 80 units. Generally NBR is less plastic than NR under mixing conditions. It may show more heat build up during mixing under identical conditions as those of NR mixing.

Copolymers containing acrylonitrile and isoprene (NIR) are also prepared. They may be easier to plasticize than NBR.

A very important characteristic of NBR is the poor solubility and diffusibility of sulphur in this rubber. This can be taken care of, by adding sulphur in the beginning of the mixing cycle, unlike other rubbers. Cure systems contain not more than 1.5-2 phr of sulphur. Since unsaturation levels are lower than those of NR, curing will be slower than NR. Peroxide cure is also done in applications where better resistance to compression set is required. Peroxide cure also gives better surface finish as in case of sulphur cure, it may bloom to the surface and spoil the appearance.

Due to its polarity, NBR may band well with the mill roll but it has poor tack – for products like hoses, tackifier may be necessary in the compounds. NIR has better tack. Plasticisers may be mostly ester types as mineral oils are less polar and hence may be less compatible with this rubber. Among the carbon blacks, NBR does not find applications which demand high mechanical properties like tyres and hence most often, formulations may include only less reinforcing fillers like SRF or FEF of GPF blacks. In fact, for the same level of hardness needed, it will be better to add SRF in higher loadings than HAF at lower loadings because, higher black loading means lesser volume of rubber and hence lower swelling in the solvent – after all it is the rubber which is affected by the solvent (swells) with which the product may be in contact, and not the carbon black.

Applications: Most of the products made by NBR are due to its excellent solvent and oil resistance (non polar). Thus, many automotive components are made of NBR especially hoses and oil seals. NBR can withstand temperatures upto 120°C for continuous use.

Other products include automotive transmission seals, V belts, synthetic leather, printers' rollers, cable jacketing.

NBR latex can be used to prepare adhesives, impregnation of paper, binder of pigments etc.

Carboxylated nitrile rubber: This is produced by copolymerization of butadiene, acylonitrile and a small amount of acrylic acid. The polymers may contain about 1 COOH group per 100–200 carbon atoms in the chain. They can be cured by sulphur and divalent metal oxides like ZnO, CaO, PbO etc. These crosslinks formed through COOH by metal ions are ionic in nature and they are heat-fugitive. Compared with NBR, it gives better mechanical properties like tensile strength and abrasion resistance. They also have better extrusion characteristics.

Liquid nitrile rubbers: Liquid nitrile rubbers can be used as reactive plasticizers for NBR – they plasticize the rubber without reducing hardness because they become part of the vulcanized network on cure. They also improve tack of NBR compounds.

Copolymers of butadiene and acrylonitrile containing terminal COOH or amine groups (called CTBN – Carboxyl Terminated Acrylonitrile Butadiene copolymer and ATBN – Amine Terminated Butadiene Acrylonitrile copolymer respectively) were developed for modification of impact properties of epoxy resins. They are added to about 5–15% by weight of epoxy resins – such combinations are used in composites and adhesive formulations. At the other extreme, epoxy resin may be added at 5–25% by weight of CTBN and such rubber formulations can undergo chain extension with cure – they may give somewhat higher mechanical properties than what pure liquid NBR can do, but still they are well short of the elasticity and mechanical properties of cured solid rubbers – still they may find uses in made-in-place sealants, castable rubber formulations etc.

Nitrile rubber-PVC blends: As we may have seen by now, NBR has many unsaturation sites and hence poor ozone resistance and also poor flame resistance. These are sought to be improved by blending with PVC. PVC also improves extrusion characteristics. For PVC, NBR acts as a permanent plasticizer and impact modifier. The blending may be done in Banbury type mixers but the important condition is that the viscosities of the two polymers must be brought to equal levels before they can blend well – this is done by fluxing at a high temperature – working up the polymers for a few minutes, together in a Banbury.

These blends are used to make LPG tubes, and other products which need better ozone resistance and flame resistance.

Neoprene (Polychloroprene): Polychloroprene rubber (CR) was manufactured even before NBR was developed. Its importance was felt soon after it was discovered and it continues to occupy an important place in rubber industry. The name neoprene was coined by DuPont. There are very few manufacturers of this rubber. This rubber is a little expensive and hence used only where absolutely essential.

This was the first rubber to be manufactured with flame, ozone and weather resistance. This was the second solvent resistant rubber developed though in many uses where solvent resistance alone was enough, NBR superceded it, due to its better solvent resistance and lower cost. Later on, where only better weather and ozone resistance were the requirements, it is superceded by EPDM. Still the market for CR continues to be considerable.

CR is manufactured by free radical emulsion polymerization of chloroprene (2 methyl butadiene), with potassium persulphate initiator. The polymerization may be stopped at 91% conversion by adding Tetra Ethyl Thiuram Disulphide (TETD). In CR, the cis polymer has higher Tg than the trans polymer (this may be due to many head-to-head links present unlike NR where only heat to tail links are present). The Tg of the high *trans* CR is about –45°C. The high *trans* content makes the chain structure fairly regular and this causes crystallization. This is advantageous for adhesive compositions – they need fast crystallization – these grades (A grades– adhesive grades) can crystallize in a few hours after milling. In strain induced crystallization, CR is similar to NR.

cis
Polychloroprene

trans
Polychloroprene

3, 4
Polychloroprene

1, 2
Polychloroprene

Structure depends on polymerization temperature. Lower the temperature more will be the crystallinity while at higher temperature more branches will form which leads to lesser crystallinity. Thus, for the usual rubber products the grades are obtained by polymerization with higher temperature while the "cold" rubbers are used for adhesives. For the usual rubber products, crystallisation needs to be controlled and hence the rubber is modified. The earlier modification was by adding sulphur. These grades are called G types. Adding sulphur during polymerization causes it to form copolymers with sulphur. Adding TETD will cleave the polysulphide bonds formed. The TETD penetrates the rubber particles and peptises the rubber. These grades will have one sulphur atom per every 100 monomer units. Thus, degree of polymerization between the polysulphide bonds is about 500. This improves processibility.

Other modifiers are mercaptans (W type). They crystallize more slowly than A types, but faster than G types.

G type CRs have a lower "nerve" on processing, good tack, high resilience, good tear strength and resistance to stress cracking. Their storage stabilities are poor.

W types have better storage stabilities, better heat and set resistance, easier to mix, but not peptisable, and higher green strength.

Later on, grades of another type called "T" type CRs have been introduced – they have still better processibility than W types, as they have some crosslinks but in mechanical properties, they are as good as W type grades. These are suitable for extrusion of profiles.

Curing: The double bonds in the *trans* structure (which is about 85% of the rubber) are deactivated by the Cl attached to them and hence the double bonds are not available for curing. It is also not possible to remove the Cl atoms attached to these bonds as they are very tightly held to the carbon atoms. Luckily there are a few units having 1,2 and 3,4 structures where the Cl atoms are linked to single bonded carbon atoms which can be easily removed by bases and through this reaction curing becomes possible.

Thus, the curatives are mainly ZnO. The crosslinking occurs through C–O–C bonds which form by removing the Cl atoms. However, the byproduct which forms is $ZnCl_2$ is acidic and this may scorch the rubber. Hence it must be removed by an acid acceptor which is mostly MgO (for superior water resistance we may use PbO). Sulphur modified grades cure fast by ZnO. This can be further speeded up by accelerator which is ethylene thio urea (also called mercapto imidazoline). W and T types need accelerator for metal oxide cure. ZnO improves heat resistance while MgO reduces it. MgO also improves plasticity.

Properties: Two things are important in CR. One is the polar nature along with the Cl atom present in each monomer unit and the other one is its crystallization behaviour. These two combined make this rubber denser than other rubbers (specific gravity about 1.25).

Since the double bonds are deactivated, the rubber has very good ageing, weather and ozone resistance. The Cl is responsible for its high flame resistance (it burns only when the flame is present while, it is extinguished when the flame is removed. It is a fairly polar polymer and hence will have solvent resistance (though not comparable with NBR). It retains mechanical properties better than NBR when in contact with non polar solvents/oils. Its permeability to air/gas is a little more than that of IIR but lower than NBR. It may not resist oxidizing acids like sulphuric and nitric acids.

When blended with NR or SBR or NBR, its curing is independent of the sulphur cure of these diene rubbers and hence cure compatibility with these rubbers is good. Blending with these rubbers is easy (when compared with the poor compatibility of NR-NBR blending). Being a polar polymer, its electrical properties are inferior and hence maybe used only for medium voltage applications.

Mechanical properties are very good and they can show high tensile strength without reinforcing fillers.

Among the sulphur modified grades, the G type grades are meant for general uses. Within these grades, a special grade known as GBRT grade has the best tack. The FB grade has lower molecular weight – this can be used for adhesives, coatings and binders. They can be modified by resins to give hard compositions (CR cannot form ebonites).

KNR grade can also be used for coatings. CG grade is a cold polymer and is used in adhesives and coatings.

Among the mercaptan modified types, FC grades are similar to FB but do not reduce scorch time greatly like FB grade. FM grades are liquids and can be used as softeners for high molecular weight grades. W grades are for general purposes. WHV grade had high molecular weight and used for oil extended and highly filled products. WHV-100 has lower molecular weight than WHV grade and better processible than WHV grade. WX grade has higher tack. W-M1 is low viscosity type and can give smooth surface finish for extrudates. Mechanical properties slightly poorer than W type. WB is better for extrusion containing gels. Strength is poorer. WRT has better low temperature resistance. S grade does not crystallize and contains cross links and hence used only for crepe soles with oil resistance. Similarly ILA grade does not crystallize and it is a copolymer containing acrylonitrile. It gives better oil resistance. It also resists higher temperatures and Freons.

T grades are easily processible while retaining mechanical properties. They can be extruded 30% faster and can be mixed 10% faster. They have lower nerve. These grades are cold polymerized and have lesser tendency to crystallize due to introduction of a few cross links. They are more easily processible than W grades but with similar mechanical properties.

A grades are fit for adhesive uses. AC and AD grades crystallize rapidly. They contain no modifier. AF and AG types do not show crystallizing tendency.

Uses: The major uses are in adhesives, transportation, energy and construction industries.

Their adhesive uses are due to their polarity and crystallinity. The polarity gives greater versatility in bonding a wide range of substrates while the crystallinity gives better strength of the bonds. Footwear industry uses CR adhesives for bonding NR or PU soles to leather uppers etc. Aircraft, automobile and construction industry used CR adhesives.

Adhesives can be based on solvents or latex. Dry film adhesives are also used in some places – they do not cure and give strength by crystallization on cooling.

In transportation industry, CR is used for V belts, timing belts, blown sponge gaskets for door, deck and trunk, spark plug boots, power brake bellows, suspension joint seals, ignition wire jackets. In aviation industry CR is used in mountings, wire and cable jackets, seals and de-icers. In railways, it can be used for track mountings, air brake hoses, flexible car connectors, freight-car interior linings and journal box lubricating pads etc.

In energy industry it has applications in exploration, production and distribution of petroleum like packers, seals, hoses coated fabrics, wires and cables.

In construction industry, the uses are window wall sealing gaskets, sheetings for waterproofing, roof covering, highway joint seals, bridge bearing pads, soil pipe gaskets. CR modified asphalt is also useful in road pavements, and in airport runways.

Wires and cables use CR extensively, mainly for jackets – the main property are ozone, weather and flame resistance. Hoses requiring flame resistance also use CR. The weather resistance is exploited in products like V belts, transmission belts and conveyor belts. Cushions based on CR are used when flame resistance is a critical requirement.

Recently due to the concerns expressed about handling chlorine, production of PVC and CR are strongly discouraged in advanced countries. This places new challenges in finding alternatives for such a highly useful rubber.

Chlorosulphonated polyethylene (CSM): We have already seen that disrupting the crystallinity in PE can give an ethylene based rubber and further, that this can be done by copolymerization with propylene. Copolymerisation with polar monomers can also achieve the same effect. There are two more approaches to obtain ethylene based rubbers. One of them is to chemically modify the polymer so that bulky side chains can be introduced into the polymer backbone in a random fashion. This can reduce crystallinity.

Two modifications are well known – one of them is chlorosulphonation and the other one is chlorination. Chlorosulphonated polyethylene is also called hypalon. It was introduced in 1952.

This polymer can be produced by dissolving LDPE in boiling CCl_4 and passing SO_2 and Cl_2. These gases react together to give sulphuryl chloride which can react with the polymer in presence of light or peroxide and add chlorine or SO_2Cl side chains through free radical mechanism. Like CR this polymer is also available as chips, white or cream in colour. The PE used may be having molecular weight of about 20000. There are various grades of CSM – the chlorine content varies in them. If the Cl content is about 30%, there are approx. 17 Cl atoms per 100 main chain carbon atoms. Sulphur content is about 1–2% in general – about one chlorosulphonic group per 150 chain backbone atoms. Chlorine content can vary between 23 and 43% and with this specific gravity varies from 1.09 to 1.27.

$$+CH_2CH_2+_n \xrightarrow{SO_2Cl_2} +CH_2CH_2+_x +CH_2-CH+_y +CH_2-CH+_z$$
$$\underset{Cl}{|} \qquad \underset{SO_2Cl}{|}$$

$z = 17$, $x + y = 12$ for common grades

The commercial grades are designated as Hypalon 20, 30, 40, 4085, 45, and 48. In these Cl contents are 29, 43, 35, 36, 23 and 43 respectively.

The grades 45 and 48 are more thermoplastic than other types. The grades 40 and 4085 give the best mechanical properties while those with numbers 23 and 30 are readily soluble in toluene-ketone mixtures. Hypalon 623 is a lower viscosity analogue or hypalon 45.

Mooney viscosities may vary from 28 to 98 at 100°C. They may withstand temperatures from about –40°C to 160°C. Continuous use temperature is 120°C. As chlorine content increases, stiffness of the chains increase but rubberiness also improves. This statement may sound strange but it must be remembered that while developing ethylene based rubbers, we reduce crystallinity and for this, we add bulky side chains – this will not only stiffen the chains but also make it more rubbery.

It is similar to EPDM in some ways – the differences being in increased polarity due to increase in chlorine content – this modifies solvent and flame resistance. It is similar to EPDM in heat, ozone, weather resistances, colour retention. Too much chlorine content will make this polymer too stiff to be a rubber.

As chlorine content increases, Tg will increase, polarity will increase and hence solvent resistance. Flame resistance will also increase. The less chlorinated grades may show a little crystallinity – hypalon –40 may show good physical properties even without curing.

Often fillers may not add to strength for these vulcanisates.

While processing these elastomers may show thermoplastic behaviour. In banbury mixer, upside down mixing may be preferred. In mill mixing, longer mixing times may not reduce properties but scorch problems may arise.

Calendering temperatures may vary – those with more crystallinity may be calendered at higher temperatures i.e. 100–150°C while other grades like hypalon–40 may be calendered at 60–100°C. The polymer is also easy to extrude or mould (compression or injection).

Curing may be done by many systems – mainly MgO and PbO. They may liberate chlorine and forms double bonds in the main chains which can be cured by sulphur and accelerator. Penta erythritol may also cause formation of additional cross links. Metal oxides may react with SO_2Cl and form disulphone ester cross link through penta erythritol. Other cure systems are epoxy resin, aliphatic or aromatic diamines and peroxides.

A unique cure system is ionic cures. This is possible by reaction of SO_2Cl with metal oxide giving ionic cross links – they are heat fugitive – disappear on heating and reappear on cooling. They may proceed at room temperature and in presence of a trace of moisture. Thus a pond liner can be cured by simply installing the calendered sheet and allowing the cure to occur on its own in presence of atmospheric moisture.

Applications: Hypalon is characterized by its combination of heat and solvent resistance. It performs better in corrosive environments, than NBR and CR. It resists acids and oxidizing agents better than NBR and CR. It is tougher than EPDM and silicones. Other important properties are colour retention and radiation resistance (hence used widely in nuclear plant insulations).

High chlorine grades have lesser resilience and hence used for damping of vibrations. Due to its favourable combination of properties, it is very much used as cable sheathing material. Its acid resistance is very good. So, it can be use for acid carrying hoses. Other uses are industrial paints (retention of light colours lead to lesser radiation absorption from sunlight and hence cost of air conditioning is reduced. It will not darken as its light colour retention is excellent. Sealing using CSM filled with magnetic fillers is another common use in refrigerators.

Chlorinated Polyethylene (CM): $+CH_2CH_2\!+\!+CH_2CH\!+$
$$\qquad\qquad\qquad\qquad\qquad\qquad\qquad\qquad | \atop Cl$$

Another modification of polyethylene is chlorination. In many ways, properties of CSM and CM are similar. The effect of increase in chlorine content in the polymer will be similar to that in CSM. Preparation of CM is by suspending polyethylene in a suitable medium and passing Cl_2 at elevated temperatures. Chlorine contents may vary from 25 to 45%. Grades with 25% chlorine are partly crystalline and hard. The best rubbery characteristics are obtained with Cl content at 35%. Grades having Cl content beyond 45% are brittle. Dow Corning produces CM in a number of grades – CM 0136 and CM 0236 both have Cl content of 36% and specific gravity of 1.16 while CM 034 has specific gravity of 1.25 with Cl content 42%. CM 0342 has better flame resistance and is less permeable to gases while CM 0236 has better water resistance.

This polymer needs stabilization against HCl evolution and this is achieved by adding MgO in the compounds. Heat resistance may be improved by adding epoxy resins and anti oxidants.

Curing is done by peroxide or radiation. Acidic substances and highly unsaturated substances may interfere with cure and hence should be avoided in the compounds. In mixing, the behaviour is similar to hypalon. In many ways it is similar to CSM except for the cure versatility of the latter. CM can be blended with other peroxide curable rubbers. EPDM blending may improve low temperature resistance while NBR makes it more oil resistance.

Main uses are in the field of cables, high pressure, heat resistant hydraulic hoses, hot conveyor belts and moulded products.

Copolymers of ethylene with methyl acrylate and vinyl acetate:

As mentioned earlier, copolymers of ethylene with polar monomers like vinyl acetate and methyl acrylate are very useful elastomeric materials. Of these, ethylene vinyl acetate (EVA) is a unique polymer which finds its place in all the three classes of commercial polymers – elastomers, plastics and thermoplastic elastomers, while ethylene-methyl acrylate (EMA) is often considered under poly acrylate rubbers rather than an ethylene based rubber.

EVA elastomers: $-(CH_2CH_2)-(CH_2CH)-$
$$\underset{\underset{OCO\,CH_3}{|}}{}$$

EVA can be obtained by free radical or anionic polymerization of ethylene and vinyl acetate. The properties depend on vinyl acetate content in the grade. Low vinyl acetate grades (less than 25%) will be like thermoplastics – the more the ethylene content, the more the polymer will resemble PE. The medium vinyl acetate grades (40–60%) can show elastomeric behaviour, while the grades having still more vinyl acetate will not be useful as elastomers. The elastomeric behaviour is due to the $-OCOCH_3$ side chain which prevents crystallization of the polymer.

Polymerization:

Properties: The copolymer formed between ethylene and vinyl acetate is a random one. Hence the polymer shows no crystallization except for the very low vinyl acetate content grades. The brittle point of the polymer is about $-50°C$ which is higher than EPDM. The specific gravity is about 0.98. Mooney viscosity values are fairly low – for most of the grades it is less than 55. The polymer is fully saturated and hence its heat resistance, ozone resistance and weather resistance are very good – better than NBR though the solvent resistance is poorer than NBR and CR (inspite of its polarity). This polymer hence will combine heat and solvent resistance. It has unlimited life at $120°C$ while about 1 year at $150°C$ and several weeks at $180–200°C$. As far as heat resistance is concerned, the polymer is surpassed only by silicones and fluorocarbon rubbers and in absence of air, it is claimed to be better than silicone. The polymer resists hot water and super heated steam, but attacked by saturated steam and concentrated acids and alkalies.

This elastomer is often available as granules and has thermoplastic nature while mixing-hence often they can be compounded in thermoplastic mixing equipment at a high production rate. Some of the grades do not require curing though peroxide cures do give better set and heat resistance.

The mechanical properties are good in presence of reinforcing fillers – elongation at break can be high but resilience is not so high unless the cross link density is sufficiently high. Abrasion resistance is not so good and tear strength is not so high – this can pose problems while demoulding.

The polymer itself does not contain toxic ingredients while manufacturing and if cured properly, can give food grade products easily i.e., safe products for contact with food materials.

Curing is effected by peroxides. Peroxide – triallyl cyanurate (TAC-co agent) combinations are used. The state of cure can be increased by co agent content rather than the peroxide content in the mix. Curing temperatures are high – often 170°C–200°C.

Electrical properties are typical of polar polymers – this can be classified under 'semi conducting' polymers – i.e. those with moderate values of resistivity.

Other useful properties are – environmental stress cracking resistance and flexibility – this makes it a possible replacement for PE in some plastic products.

Applications: In the elastomer field, the uses of EVA are rare – mainly in the field of cable insulations (for medium voltages). This polymer while burning does not release toxic gases and smoke – hence this polymer is extensively used as an insulating material for underground railway stations or high rise buildings. It must be remembered that in such places, in the event of fire accidents, more deaths are caused by the suffocation rather than burning itself – this is an area where PVC in spite of its superior flame resistance is found wanting – it releases chlorine and HCl in the event of a fire accident while EVA does not release poisonous gases or dense smoke. Though EVA does not have flame resistance, this property can to some extent be improved by flame retardant additives.

The other elastomeric application where this polymer finds a prominent place is the Hawaii chappal soles. This polymer can be mixed in thermoplastic compounding equipment and can be injection moulded like a plastic and can also give bright colours and attractive surface finish – in all these areas it is much better than the traditional NR. Hence the uses of EVA in hawaii chappals is increasing.

Other uses are in hoses and tubes for consumer products where it has much better flexibility and stress cracking resistance than PE or PP. It can also be used for packaging rubber compounding ingredients for direct feeding into Banbury mixers – it melts during mixing and is more compatible than the traditional PE – hence this is more useful in this area.

Another important area is impact modification of PVC-EVA can be a permanent plasticiser for PVC like NBR – the blending can be done easier and many flexible PVC products especially for contact with food have EVA as the impact modifier than the plasticizer which was commonly used earlier.

One of the emerging areas in the field of adhesives is the hot melt adhesive. They solve the environmental problems caused by solvents. Further in automobile industry, where assemblies are to be done rapidly, hot melt adhesives are more productive than the traditional solvent based ones because the bond strength is obtained by cooling of the melt which is much faster than solvent evaporation. For this application, a polymer must be thermoplastic, elastomeric and should give good strength for the bond after cooling. EVA fulfils all these requirements much better than its rivals like SBS block copolymers and hence a preferred polymer in this industry. Further EVA is a low cost polymer unlike SBS. Thus this has become a major application for EVA over the years.

Other uses may be in the field of roofing sheets, medical goods, asphalt modification etc.

Poly acrylate rubber (ACM):

The polymers of acrylic esters are polar and contain no unsaturation and hence may be expected to exhibit good solvent with heat resistance.

The first acrylic rubbers were introduced by Goodrich in the 40s. The initial rubbers had ethyl acrylate and butyl acrylate as monomers. The Tg values of poly ethyl acrylate and poly butyl acrylate are $-22°C$ and $-56°C$ respectively. As the length of the alkyl side chain in the ester group increases, Tg may decrease but solvent resistance in the resulting polymer may also decrease. Hence an acrylic rubber may be made by polymerization of ethyl acrylate with a monomer which may help in reducing the Tg (hence methoxy ethyl or ethoxy ethyl acrylate) together with a third (cure site) monomer. Alternatively, it may be prepared by polymerization of butyl acrylate with acrylonitrile (for improving oil resistance) with a cure site monomer. The cure site monomer may be 5% by weight. The cure site monomers may be 2 chloroethyl vinyl ether or vinyl chloro acetate or acrylic acid or allyl glycidyl ether (this has an epoxy ring which can take part in cross linking by ring opening). A few self curing blends are also available which on blending and heating will cure without any other reagent being added.

Main monomers :

$$+CH_2CH+ +CH_2-CH+ \qquad +CH_2-CH+$$
$$\quad | \qquad\qquad | $$
$$COOC_2H_5 \quad COOBu \qquad\qquad COOCH_2CH_2OCH_3$$

Ethyl acrylate   butyl acrylate   methoxy ethyl
                                         acrylate

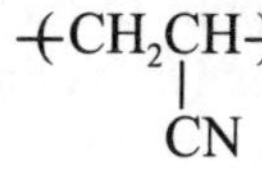

$$+CH_2CH+$$
$$|$$
$$CN \quad ACN$$

Cure site monomers :

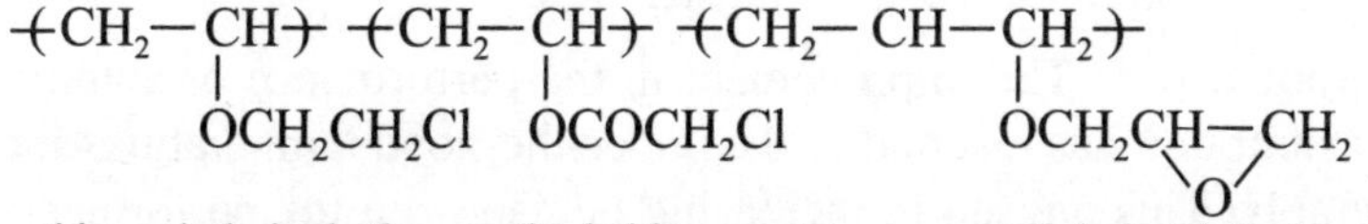

$$+CH_2-CH+ \quad +CH_2-CH+ \quad +CH_2-CH-CH_2+$$
$$| \qquad\qquad | \qquad\qquad\qquad |$$
$$OCH_2CH_2Cl \quad OCOCH_2Cl \qquad OCH_2CH-CH_2$$
$$\qquad\qquad\qquad\qquad\qquad\qquad\qquad\qquad \backslash O /$$

chloroethyl vinyl ether   vinyl chloro acetate   allyl glycidyl ether

The polymerization is done by free radical emulsion polymerization. The copolymers are of random type and they do not crystallize on stressing. Their specific gravity is about 1.1. They have thermoplastic behaviour during mixing and some times upside down mixing procedure is preferred in internal mixers.

They calender and extrude smoothly. They need reinforcing fillers for optimum mechanical properties.

Curing: Cure system depends on the grade used. This is because, each grade will have a different cure site monomer and hence cure system will depend on this. Hence, the suppliers' instructions in this regard must be observed carefully.

Those which contain chloroethyl vinyl ether as the cure site monomer may be cured with diamines, ethylene thio urea or poly amines. Addition of sulphur and metal oxide improves heat resistance and makes the formulation scorch resistant. Examples for such systems are triethylene tetra amine with sulphur and MBTS, hexamethylene diamine carbamate with dibasic lead phosphate and ethylene thio urea with lead oxide. Those grades containing vinyl chloro acetate can be cured by the unique system – sulphur-soap system – e.g., sulphur 0.3 phr and potassium stearate 3 phr. This cure system is convenient and flexible. Increase in soap content or sulphur content will speed up the cure while an increase in soap level alone can improve ageing resistance. The resulting cross link has C–O–C type structure. Those grades containing epoxy rings can be cured by ammonium benzoate or adipate. Often ACMs require post curing in an oven after the press cure, to optimize the properties.

Properties: The heat resistance is upto 200°C and in the lower side they can withstand down to –25°C. They can resist oxidation even at high temperatures. They also have good ozone, weather and flex resistances.

Besides these, they also withstand solvents and oils at such high temperatures. Another important advantage is that they can withstand oils and fuels which may contain sulphur – in case of NBR sulphur in fuel can be absorbed by the rubber (NBR) in contact and react slowly, leading to embrittlement – ACM does not suffer such disadvantages.

They also have low permeability to gases and retain light colours very well. Further, they do not give out toxic fumes when they burn. All these properties are exploited by the industry.

Applications: The improvement in the performance of automobiles in fuel consumption has resulted in the under-the-hood temperatures increasing considerably. This has led to increasing demands on the performance of the automotive rubber components – mainly hoses, seals etc. Thus components which require enhanced heat resistance can be made of ACM instead of NBR. Transmission seals, crank shaft seals, coatings, adhesives, beltings, hoses, etc., are some of the products made of ACM. ACMs also have vibration damping characteristics and hence may be used to make vibration damping mounts.

Ethylene methyl acrylate (EMA or Vamac) rubbers: This also can be considered as an acrylate rubber. This is made by copolymerizing ethylene and methyl acrylate by solution polymerization. A cure site monomer is also present – acrylic acid.

$$-(CH_2CH_2)--(CH_2CH)--(CH_2CH)-$$
$$COOCH_2 \quad COOH$$

The Tg values of this polymer will depend on the methyl acrylate content. For example a 50:50 copolymer will have a Tg of about –46°C. With increase in methyl acrylate content, the polymer will become progressively more polar and more oil resistant. With more ethylene, better mechanical properties will be observed.

During processing this polymer too shows thermoplastic character. Curing can be done using diamines or peroxides. Often, methylene dianiline or hexamethylene diamine are used for curing. Accelerator may be a guanidine – ZnO should be avoided due to its high reactivity to COOH group present in the cure site monomer.

In oil resistance, this polymer will be poorer than NBR but comparable to CR. Its upper service temperature range may be upto 200°C.

ACMs are not recommended for applications where contact with steam or acid or alkali are possible – to some extent EMA can withstand these environments.

Poly ethers and Poly epichlorohydrins:

Epichlorohydrin is an important chemical used for producing epoxy resins which are very important thermosetting plastics. Later, this monomer has entered the elastomer industry.

These elastomers are called Hydrin elastomers. Poly epichlorohydrin is designated as CO rubber, while its copolymer with ethylene oxide is ECO and its terpolymer with ethylene oxide and allyl glycidyl ether is GECO.

$$CH_2CHCH_2-Cl-\text{epichlorohydrin} \xrightarrow{\text{Polymerize}}$$

$$(CH_2-CHO)\ (CH_2-CHO)$$
$$\quad\quad\quad |\quad\quad\quad\quad\quad\quad |$$
$$\quad\quad\quad CH_2\,Cl\quad\quad\quad\quad CH_2\,Cl$$
$$\quad\quad\quad\quad\quad\quad\quad\quad\quad\quad CO\ rubber$$

$$+CH_2CHO)-(CH_2CH_2O)-\ \text{-ECO (copolymer)}$$
$$\quad\ |$$
$$\quad CH_2Cl\quad\quad \text{ethylene oxide\ rubber}$$

$$+CH_2CHO)-(CH_2CH_2O)-(CH_2CHO)-(OCH_2CH)-\ \text{terpolymer}$$
$$\quad\ |\quad\quad\quad\quad\quad\quad\quad\quad\quad\quad |\quad\quad\quad\ |\quad\quad\quad\quad\quad\quad GECO$$
$$\quad CH_2Cl\quad\quad\quad\quad\quad\quad CH_2Cl\quad\ OCH_2CH=CH_2$$

Epichlorohydrin can be polymerized by ring-opening polymerization method, with R_3Al + water + acetyl acetone catalyst. It is similar to ionic polymrieization of butadiene and isoprene.

These polymers are some what more thermoplastic and may be processed in a manner similar to other polymers showing this character. They extrude and calender well. Some times sticking to the mill roll may be observed and hence the compound must be cooled well during mixing. Mould fouling due to the sticky and corrosive nature of poly epichlorohydrin is also a factor to be considered.

Curing may be done using metal oxide or other systems (diamine + metal oxide) useful for halogen containing polymers. Peroxide cure gives better set resistance. The terpolymer contains double bond in the side chains and hence may be cured by sulphur and accelerator apart from peroxide.

Properties: The Tg of the homopolymer is quite high (–15°C) and its resilience is low. The polymer is resilient only above 50°C. The low temperature properties may be improved by copolymerization with ethylene oxide which reduces the polar attractive forces between the chains. The terpolymer has a Tg of about –40°C. The solvent resistance of CO and ECO rubbers are comparable with NBR (with high nitrile content). These polymers resist aromatic liquids as well as NBR. Heat resistance is better than NBR but poorer than ACM. The resilience of the terpolymer is good. CO rubber has good damping properties and low gas permeability. ECO gives a favourable combinations of properties like low temperature resistance, oil and heat resistance with low hysteresis and also flame resistance – a combination cannot be offered by NBR or ACM.

A related polymer is poly propylene oxide copolymer with allyl glycidyl ether. This has much better low temperature properties than CO rubbers but poorer solvent resistance. They can be cured by sulphur and accelerator.

Uses: These polymers are used for seals, hoses, mountings etc.

Poly sulphide rubbers: These were the first solvent resistance rubbers discovered (in the 30s). They have excellent solvent and oil resistance – in these respects they are unsurpassed by any rubber. Due to their deficiencies in other aspects their market has shrunk but they still have some uses.

Polymerization is done by condensation polymerization by the reaction of alkyl dihalides and sodium poly sulphide.

$$nCl-R-Cl + nNa_2S_x \longrightarrow -[-RS_x-]_n- + 2nNaCl$$

The reaction maybe conducted with slow addition of the dichloride to a solution of sodium poly sulphide at 70°C. Magnesium hydroxide is added as a protective colloid due to which the polymer forms in a latex stage from which the rubber may be coagulated.

The dichlorides may be ethylene dichloride or dichloro propane or bis (2 chloroethyl) ether or glycerol 1,3 dichlorohydrin. The value of x may be approx. 2–4. The grades FA and A may have high molecular weights. They may be masticated with chemicals for easy processing. In the latex state, addition of sodium sulphite or hydrosulphide can cleave the bonds and reduce molecular weights. By this way it may also be possible to prepare liquid polymers (i.e., of low molecular weight). Adding a small amount of trichloro propane during polymerization leads to formation of branched polymer (Thiokol ST). Poly sulphide rubbers of grades Thiokol A and FA are terminated by OH groups which are formed by hydrolysis of terminal chlorine groups in the alkaline medium of the sodium polysulphide. When high molecular weight grades are decomposed to give liquid rubbers in latex stage, liquid rubbers with SH groups are formed (Thiokol ST grades).

Thiokol A and FA grades are plasticised by adding MBTS which cleaves some of the S-S links. The grades plasticized by this way can be cured by ZnO. The resulting cured polymer is only a high molecular weight polymer and not a cross linked one. Hence these grades show a high compression set. The mouldings must be removed from the mould only after cooling.

Polymers with SH terminal groups are cured by oxidative linking of the long chains with formation of disulphides.

$$2-R-S-H \xrightarrow{O_2} -RSSR- + H_2O$$

The oxidation may be effected by metal peroxides like zinc peroxide. Quinine dioxime may also be used for this purpose. These networks are cross linked and hence may be taken out of the mould even when hot.

Properties: These rubbers need reinforcement – 50–60 phr of FEF or SRF black may give strength. Processibility of these rubbers may be improved by 10–20 phr of CR or NBR or NR or liquid poly sulphide rubbers. They have fully saturated backbone and hence will have excellent ozone and weather resistance. The more the sulphur in the rubber, the more polar and hence more solvent resistant and less permeable to air and gas but also less resistant to low temperatures and less resilient. However, some grades do have good low temperature properties – they were formerly used for window sealing in aircraft though today, in this application, silicones have come in a big way. Thiokols FA and ST can resist even aromatic oils. They resist dilute acids but not dilute oxidizing acids.

Thiokol DA is made from bis (2 chloroethyl) ether. It has 47% sulphur content and specific gravity of 1.34 while Thiokol A has 85% sulphur (specific gravity 1.6). Thiokol ST made from bis (2 chloroethyl formal) + 1,2,3 trichloro propane has sulphur content 37% and specific gravity 1.27.

The major disadvantages are poor mechanical properties, poor heat resistance and bad odour during processing.

Applications: The main property which is exploited, is the solvent resistance. Printers' rollers, sealants, hoses, diaphragms for gas meters etc. are made of this rubber. Other uses may corrosion resistant coatings. Liquid Thiokol rubbers may be used for castings, adhesives, binders for rocket fuels, flexible moulds, sealants, modifiers for epoxy resins, electrical encapsulations etc.

5

High Performance Elastomers

These elastomers show the remarkable ability to withstand temperatures of about 250°C. The rubbers which fall in this category are the fluorocarbon and silicones. Poly urethanes are also covered in this chapter though strictly speaking they do not fall in this category - their only high performance characteristic is in terms of ability to withstand mechanical abuse (i.e. abrasion) – they are special because they can be processed by unconventional rubber processing methods.

Fluorine containing rubbers: Fluorine is a very unique element. On one hand it is highly electro negative and on the other, its atomic size the second lowest, next only to hydrogen. Hence, organo fluorine compounds will have much stronger inter molecular attractive forces compared with the similar chlorine compounds. The C–F bond energy values are very high and this coupled with its small size make perfluorinated chains possible – it is to be noted that the stabilities of perchlorinated higher alkanes are very limited. Extending this logic further, we can guess that polymers containing fluorine alone in the side chains are possible while this is not possible in the case of chlorine or the higher halogens. The fluorine atoms are more tightly bound to the carbon and form protective shields around the main chains and this protects the polymers from chemical attack while at the same time leading to close packing of the chains. Thus PTFE (poly tetra fluoro ethylene) will be a plastic though its Tg should be low (since the side chain is very light). The Tg value for PTFE is uncertain but said to be low.

Attempts to develop rubbers based on fluorine containing monomers failed. Many monomers like fluoroprene, vinylidene fluoride, fluoro acrylate etc., were tried but these were not satisfactory. Finally it was found that copolymers containing vinylidene fluoride (VDF) can offer satisfactory elastomeric properties. Among the fluorine containing polymers, these (VDF copolymers) are most successful. We also have rubbery derivatives of tetra fluoro ethylene which give the best possible resistance to oxidizing chemicals, the nitroso fluoro rubber and phospho nitrilic fluoride rubbers which are the best possible rubbers for flame resistance, besides fluoro silicone rubbers which combine the advantages of fluoro carbon rubbers and silicones.

VDF based rubbers (FKM): PTFE was discovered in 1938 and found to be highly inert and the most slippery solid substance. It was also found that VDF was leathery a few copolymers based on VDF can be rubbery. In the 50s, the aerospace industry needed elastomers with better heat and fuel resistances for military jet engines. For this copolymers based on VDF were tried, the first one being the one with CTFE (chloro trifluoro ethylene). Soon the better one, i.e. VDF + HFP (hexa fluoro propylene) was introduced – this copolymer is called Viton A by Du Pont. A still better grade with best solvent resistance i.e. a terpolymer containing VDF, HFP and tetra fluoro ethylene (TFE) – called Viton B was also introduced.

Later, the applications spread to civilian areas too. The processing problems associated with this rubber was later solved by the use of delayed action amine cure systems.

Preparation of these polymers is by free radical polymerization with persulphate initiator in emulsion medium. The monomers must be pure otherwise the polymerization will be inhibited. The emulsifier should be a perfluorinated one otherwise chain transfer will take place. The polymerization should be done with care otherwise explosion may be possible.

The polymer is quite tough. It is soluble in some ketones, esters etc. Viton A has Tg of about –20°C, while Viton B has Tg of about 0°C. While for other rubbers brittle point is above Tg, for vitons it is below Tg – thus they can be used as rubbers even below their Tgs. Viton A may contain vinylidene fluoride upto 65%. In KEL-F, VDF content may be 50% in KEL-F 5500 grade while in KEL-F 7000 grade it may be 70%. The grade ECD 006 which contains per fluoro methyl vinyl ether, is soluble only in highly fluorinated solvents.

The copolymers (FKMs) have the following compositions: Viton A: VDF + HFP (hexa fluoro propylene), Technoflon SL and SH: VDF + HPFP (hydro penta fluoro propylene), Viton B:VDF + HFP + TFE, Technoflon T:VDF + HFPF +TFE, ECD 006: TFE + X + PFMVE (perfluoro methyl vinyl ether), while copolymer of the type CFM has grades called KEL – F 3700 or 5500 or SKF-32: VDF + CTFE (chloro trifluoro ethylene). In ECD 006 grade, PFMVE is 15%, TFE 10% and VDF 75% – Tg is about –37°C and brittle point –50°C.

Molecular weights may vary. In viton LM grade it is 5000, in viton A it is 1,00,000, for Viton AHV, it is 200,000 and in KEL-F it is several lakhs. More fluorine content, leads to better thermal stability while more H leads to more reactivity.

Viton A $+CF_2CH_2+$ $+CF_2CF+$
$\qquad\qquad\qquad\qquad\qquad$ |
$\qquad\qquad\qquad\qquad\quad$ CF_3

Viton B $+CF_2CH++CF_2CF++CF_2CF_2+$
$\qquad\qquad\qquad\qquad\quad$ |
$\qquad\qquad\qquad\qquad$ CF_3

Technoflon SL $+CF_2CH_2++CF_2CF+$
$\qquad\qquad\qquad\qquad\qquad$ |
$\qquad\qquad\qquad\qquad\quad$ CF_2H

Technoflon T $+CF_2CH_2++CF_2CF++CF_2CF_2+$
$\qquad\qquad\qquad\qquad\qquad$ |
$\qquad\qquad\qquad\qquad\quad$ CF_2H

KEL $+CH_2CH_2+$ CF_2CF
$\qquad\qquad\qquad\qquad\quad$ |
$\qquad\qquad\qquad\qquad\quad$ Cl

ECD006 :
$(CF_2CF)(CF_2CF)(CF_2CH_2)$
$\quad$ |
$\quad$ OCF_3

These polymers may be cured by diamines + metal oxides. The cure may proceed in two stages – first, press cure at 150–180°C for about 10 mins – this is to be followed by post cure in oven for 10–24 hrs at 260°C to optimize the set and other properties. Bisphenol-A can also cure this polymer – this gives better set resistance and better scorch safety.

The polymers may pose problems in mixing – to some extent this is solved by adding a small amount of low molecular weight polymer of the same type during mixing.

Properties: These polymers are quite stiff and have good strength even without fillers. The strength is high because at room temperature, they are only 40°C above the Tg. Their elongation at break values are quite low – about 250%. Set resistance is good even after long exposure to 200°C. The life of Viton products is high – at 200°C is infinite while at 230°C it is 300 hrs and at 290°C it is 100 hrs and at 315°C it is 48 hrs. Their low temperature properties are poorer. Lower service temperature may be around –10°C.

Their solvent resistance is excellent – only some ketones may swell them. Their ozone weather, abrasion, steam and tear resistances are exceptionally high. Their resilience is low at room temperature.

KEL rubbers have better low temperature resistance than vitons though their heat resistances are poorer. Their main property is high resistance to fuming nitric acid.

Technoflon rubbers are poorer in heat resistance due to their lesser F content. ECD 006 rubbers have heat resistance even better than vitons. Even at 260°C their compression set is much lower than other rubbers.

Uses: Mainly for highly demanding applications – due to their high cost. – gaskets for aeronautical and automobile industry, hoses, membranes, non flammable coatings on flammable materials, binders for asbestos etc.

KALREZ Rubber: This is a perfluoro rubber containing perfluoro methyl vinyl ether, TFE and a cure site monomer which also contains fluorine mostly except for the active element or group. Curing is done by diamine + metal oxide. It is a rubbery derivative of TFE. It is the ultimate in chemical and heat resistance. It combines the superior heat and chemical resistance of PTFE with resilience of a rubber. It resists almost all chemicals except molten sodium or potassium metals. It can withstand 290°C, continuously and 316°C intermittently. Processing, this is very difficult and requires specialized tooling and hence it is directly sold as products like rods or seals etc.

$$-(CF_2CF)-(CF_2CF_2)-(O-\underset{F\quad F}{\overset{F\quad F}{\bigcirc}}-F)$$
$$\quad\quad\;| $$
$$\quad\quad OCF_3$$

Curing may be through
elimination of this F atom

Nitroso rubber: This rubber contains N-O bonds in the main chain apart from PTFE segments. This is non flammable (even in a pure oxygen atmosphere) rubber and hence it was studied extensively. However, due to drawbacks like difficulty in synthesis and cross linking, poor mechanical properties, toxic decomposition, poor heat resistance (relative to other fluoro rubbers) and very high cost, it is not much of a commercial success.

Structure of the polymer is: $-(CF_2-CF_2)_n-N-O$
$$\quad\quad\quad\quad\quad\quad\quad\quad\quad\quad\quad\quad\quad\quad\quad\quad |$$
$$\quad\quad\quad\quad\quad\quad\quad\quad\quad\quad\quad\quad\quad\quad\quad CF_3$$

Curing may be done by diamine + metal oxide but this gives poor mechanical properties. Hence a ter polymer containing –COOH side chain has been developed which can be cured by metal complex.

Phosphonitrilic fluoro rubbers (PNF rubber): This rubber contain P = N as the main chain. This can be called an inorganic polymer as its main chain does not contain carbon. The precursor of this rubber is the inorganic polymer $-(PNCl_2^-)_n-$. This is prepared by reaction of ammonium chloride and phosphorous penta chloride. This is a non flammable polymer but highly prone to hydrolysis. To avoid this, the Cl atoms in the side chains must be replaced by highly fluorinated alkyl chains. This helps in making the polymer water resistant and also to push the chains further apart from each other and thus bringing down the Tg to very low value (–68°C)

$$\begin{array}{ccc} Cl & & OCH_2CF_3 \\ | & nNaOCH_2CF_3 & | \\ -(-C=P-)_n + & + \longrightarrow & -(-N=P-)- \\ | & nNaOCH_2(CF_2)_3CF_2H & | \\ Cl & & OCH_2(CF_2)_3CF_2H \end{array}$$

This is also a non flammable rubber like nitroso rubber but more useful. Service temperature range between $-55°C$ and $+150°C$ approx. It is cured by peroxide or sulphur. It has good solvent resistance and better low temperature properties than most of the other fluorine containing rubbers but poorer in heat resistance than them.

Perfluoro alkylene triazine rubber:

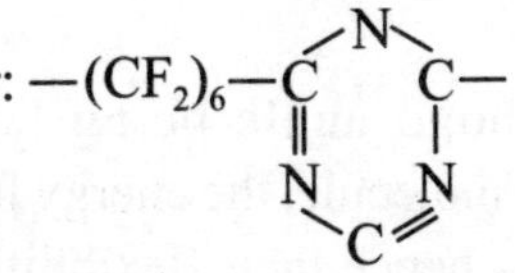

$$-(CF_2)_6-\underset{\underset{N}{\|}}{C}\overset{\overset{N}{\diagdown}}{}\underset{\underset{C}{N}}{C}-$$

This rubber has good mechanical strength and excellent heat and oxidative resistance but poor hydrolytic stability.

Fluoro silicone rubbers (FMQ):

$$-Si-O-$$
$$\overset{CH_3}{\underset{CH_2CH_2CF_3}{|}}$$

(FVQ:contains vinyl groups in place of the $CH_2CH_2CF_3$ in a few repeating units)

This combines the advantages of FKMs and silicones i.e. the solvent resistance of the former and the low temperature resistance, resilience, lower hardness and bondability to other substrates, of the latter.

Silicones: Silicones are important technically – their uses are in the fields of rubbers, resins, lubricants, vaselins, lacquers, grease, defoamers, insulators, medicines etc.

The general structure of a silicone polymer is :

If R is methyl, the polymer is called poly dimethyl siloxane

$$-(-\underset{R}{\overset{R}{|}}\underset{|}{Si}-O-)_{n-}-$$

(MQ). If some methyl groups are replaced by vinyl (MVQ) and some by Ph group, MPQ and MPVQ etc.

These polymers can be called inorganic polymers because the main chain does not contain carbon. Due to the Si–O main chains they have unusual properties. One of the important structural factors of silicone polymers is the bond angle of the Si–O–Si bond which, unlike the C–C–C bond angle seen in the other polymers is not the usual tetrahedral angle (109°), but 134°. This makes the siloxane chains highly mobile. This also favours formation of large rings – this may account for formation of highly ordered structures consisting of helices rather than a random chain structure. These helical chains may be coiled into larger bulky rings of 50–100 dimethyl siloxane large rings. They may form a double helix – this structure indicates this chain will flow easily rather than form a compact mass – indeed these polymers have very low viscosities compared with other rubbers. This structure can be easily destroyed by stirring etc. – hence their die swell etc, may be excessive – they may be difficult to extrude.

With the bond angle being large, compared with the analogous polyisobutylene molecule, the energy for rotation of the main chain bonds will be much lower – hence their flexibilities are much more than other rubbers- their Tgs are much lower than other rubbers. MQ crystallizes at about –67°C while its Tg is –123°C. The cohesive forces will also be much lower – this is because, the inter chain distances will be much larger than the usual organic polymers – thus the viscosities are much lower.

On heating, the inter chain forces should weaken and flow should be easier but this is opposed by polar, attractive forces which operate when the chains come closer together and straighten – thus over a wide temperature range, the viscosity decreases little – similarly modulus of the cured rubber will be little affected even after prolonged exposure to heat. Thus the heat resistance is very high and at the same time, they cannot be extruded etc. Thus the normal rubber processing industries will find it difficult to handle this rubber – hence, often the supplier gives the polymer into which filler is already incorporated.

The bond energies of the Si–O–Si bond is much higher than that of the C–C–C bonds. This is an additional reason for the high heat resistance of this rubber – the other reason being their tendency to form ring structures due to the bond angle. Another reason suggested for the exceptional heat resistance of silicones is the absence of an energetically favourable mechanism by which a stable product (i.e. silicon dioxide) is formed on heating. The electro negativity difference between Si and O atoms is larger compared with that of C and H atoms – hence the ionic character of Si–O bond is higher – they exert more attractive forces over the C–H bonds in the side chains. Benzene rings attached to Si atom increase heat resistance – hence MPQ will have still better heat resistance. The higher polarity of Si–O bonds than C–H bond does not mean that silicone rubbers are polar and they can be expected to resist non polar solvents and oils. The methyl and other organic side chains in the more coiled structure will lead to paraffinic properties on the surface of the chains – thus the release properties are excellent – hence the extensive use of silicones as release agents in polymer processing. The surface tension is very low and this also leads to more permeability to gases.

In MPQ, the bulky phenyl side chain prevent uncoiling of the molecules on heating – this leads to more repulsive forces between the chains – hence it has better low temperature properties – can withstand even –90°C while MQ stiffens slightly at –60°C. The Ph content in such polymers is between 5 and 16mol %. Too high a Ph content makes the polymer brittle. PMQ will withstand heat, steam and corrosion better and has lower air permeability than MQ. Further, they are not affected by ionizing radiation.This property makes silicones highly useful for nuclear applications.

Replacement of a few O atoms by benzene rings reduce the chain flexibility (due to reduction of the bond angle) and increases inter chain attractions – thus, a polymer like

$$-\overset{\displaystyle |}{\underset{\displaystyle |}{Si}}-Ph-\overset{\displaystyle |}{\underset{\displaystyle |}{Si}}-O-$$ will be more rigid and have better mechanical properties – they have better heat resistance – if Ph is replaced by a carborane ring, the heat resistance will be still better – even up to 400–500°C. The carborane ring also increases autohesion and adhesion to other materials.

The organic side chains present around the main Si–O chain make the organic (non polar) characteristics dominate the polymer properties especially in areas like solvent resistance – to improve resistance to non polar solvents/ oils, polar sidechains may be incorporated – as done in the case of fluoro silicone rubbers (FVQ) – trifluoro propyl side chains are often added to increase solvent resistance while retaining the heat resistance and flexibility of the normal silicones.

Silicones have resistance to temperatures of 180°C (infinitely) and for short periods at 250°C while at the lower end, –75 to –100°C. Set resistance is high even at high or low temperatures. Exposure to such high temperatures cause little change in mechanical and electrical (dielectric strength, corona resistance and) properties. Oxygen, Ozone and weather resistances are also excellent. Water resistance is also excellent. It is neither corroded nor does it corrode any material in contact with it. More importantly it is resistant to biological fluids and hence its wide usage in medical applications especially implants. It repels molten polymers, but can be bonded to metals, glass etc. It has no smell. It burns without releasing much smoke and leaves behind a residue (silica) after burning-hence airport insulations are always based on silicones as in the event of a fire, the residue around the conductor ensures that communication breakdowns will not occur. Curing may be done without much pressures and even at room temperatures. Liquid silicones can undergo chain extension in addition to cross linking leading to mechanical properties achievable even from solid rubbers. Thus its uses are wide – in the areas of electrical industry, construction, food, chemical, medical, metallurgical industry.

Synthesis: The key to produce silicone rubbers is to prepare dichloro silanes:

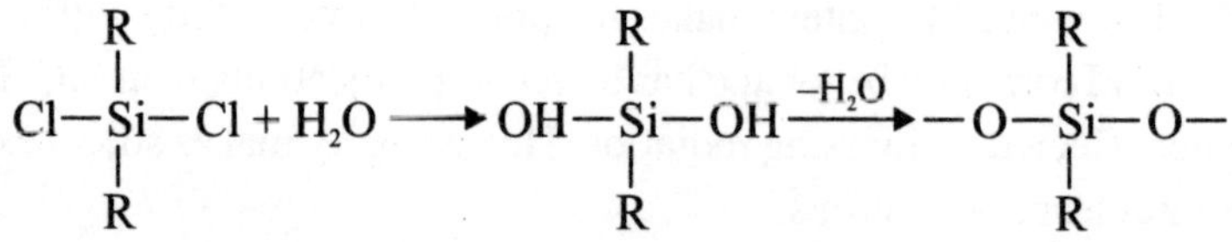

$$\underset{R}{\overset{R}{Cl-Si-Cl}} + H_2O \longrightarrow \underset{R}{\overset{R}{OH-Si-OH}} \xrightarrow{-H_2O} \underset{R}{\overset{R}{-O-Si-O-}}$$

The dichlorosilanes on hydrolysis give dihydroxy silanes which are highly unstable and immediately lose water to condense and give siloxane chains (thus this is a condensation polymerization which occurs spontaneously).

The monochlorosilanes give the monohydroxy silanes which cannot polymerize while the trichloro silanes will directly give a cross linked network (this is used to prepare silicone resins which are rigid thermosets used extensively in electrical/electronic industry). Hence it is imperative that the dichloro silanes must be purified and free from contamination from mono and trichloro silanes.

Preparation of dichloro silanes:

A number of methods are available for this purpose.

1. $RCl + Si \xrightarrow[250-280°C]{Cu} R_2SiCl_2 + R_3SiCl + RSiCl_3 + HSiCl_3 + SiCl_4 + R_4Si$

2. $PhCl + Si \xrightarrow[375-425°C]{Ag} Ph_2SiCl_2 + PhSiCl_3 + SiCl_4 + PhH + Ph-Ph$

Other methods may be by Grignard reagent :

$$SiCl_4 + RMgX \longrightarrow R_nSi_{4-n} + MgX_2$$

(Grignard method is versatile but not very convenient)

The hydrolysis of R_2SiX_2 initially gives a cyclic tetramer which can be easily separated from the other components by distillation. These in turn may be polymerized by ring opening techniques in presence of strong acids or alkalies. Earlier the dichlorosilanes were directly hydrolysed. At higher temperatures, depolymerization also occurs at a considerable rate. The catalyst concentration should be kept at a minimum but this leads to very high molecular weights and hence chain transfer agents are used to control molecular weights. The chain transfer agents not only control the molecular weights but also enable introduction of the desired end groups – vinyl or alkoxy or OH etc. Thus liquid silicone rubbers are easily made and processed.

During the polymerization, some active centres remain and these form strong bonds with fillers (silica) and make compounds unprocessible after a few days of storage. So, the catalyst (often hydroxides of alkali metals) may be neutralised by acidic substances. Further, the catalyst should be active at lower temperatures but must decompose at higher temperatures so that purity of the resulting polymer can be maintained *e.g.* $R_4N^+OH^-$. For typical rubber applications, polymers with molecular weights of a few lakhs are prepared. Molecular weight distributions should be narrow as both low and high molecular weight fractions give trouble – the lower ones reduce the elasticity while higher ones reduce processibility. The low molecular weight fractions and residual monomers (they cannot be distilled off as attempts to do so may cause decomposition of the polymer) may increase shrinkage of the polymer during cure – already, the unusual bond angle of the Si–O–Si bond contributes to increased shrinkage compared with other elastomers.

Types of silicone rubbers: Poly dimethyl siloxane is used in coatings. The low molecular weight grades may contain OH groups and may be cured by silicates etc. MVQs may contain 0.2 mole % of vinyl groups – they can be cured faster and give lower set values and hence used in extrusion and moulding.

MPVQ may be used for low temperature and nuclear applications while FVQs give better solvent and low temperature resistance.

Curing: Curing is done by peroxides. The mechanism is through formation of alkyl radicals in the side chains which join each other by radical combination. The peroxides are not dispersed uniformly through the rubber and hence cross linking is not uniform – this leads to improper distribution of the cross links in the network and hence high values of set. This can be reduced by improving the dispersion of the curative-by presence of vinyl side chains (as in MVQ) at regular intervals in the chain. Further, curing is affected by oxygen strongly – hence press cure is a must. The dosage of the peroxide may be much lesser than that used in other rubbers, in the compound.

Among peroxides, acyl peroxides like benzoyl peroxide may evolve acidic byproducts which may degrade the rubber – hence 2, 4 dichloro benzoyl peroxide is better as its byproduct is not volatile and hence degradation is avoided. If vinyl groups is not present (as in MQ), acyl peroxide has to be used.

For the rubber containing vinyl groups, alkyl peroxides are suitable – like di t-butyl peroxide or dicumyl peroxide (this is preferred for carbon black filled compounds).

Radiation cure is also useful.

Liquid silicones may contain vinyl side chains which may react with hydrosilane (in another part of the rubber), in presence of catalyst at 60°C and cross link fast.

$$\underset{\underset{O}{|}}{Me-Si-H} + \underset{\underset{O}{|}}{CH_2=CH-Si-Me} \longrightarrow \underset{\underset{O}{|}}{-Si-CH_2CH_2-Si-}$$

In these cases, though the polymers are low molecular weight materials, the cross linking is accompanied by chain extension and hence the properties are comparable with those of solid rubbers.

Silicones always require post curing in oven to remove the peroxide decomposition products which will otherwise degrade the rubber (like other peroxide cured rubbers).

Further, oxygenation of the $-CH_2-$ cross link occurs, stabilizing the vulcanizate further. Post curing is done at 150°C for 2-4 hrs and then at 200°C for 16 hrs. For still higher heat resistance, further heating at 220–250°C for 2–4 hrs.

Properties: Silicone compounds require special processing steps – mixes undergo crepe hardening – due to hydrogen bonds forming between the OH group present in silica filler surface and the oxygen atom present in the silicone polymer chain. The number of such bonds increase with time – hence freshly prepared compounds will not pose problems. The crepe hardened compounds should be subjected to "freshening" operation in which the compound is repeatedly passed through a mill with decreasing nip gaps which breaks H bonds. Often the supplier gives the pre mixed compound and the moulder some times may add only the curatives in his factory or gets the compound in which the curative too has been added already. Extrusion is difficult for this rubber and special compounding procedures must be adopted for extrudable mixes. As far as compounding is concerned, different grades – pressure curable and pressure-less curing types etc, are available.

Grades: Dow Corning has a number of grades in silicones:

General purpose grades may consist of Silastic GP-30, GP-45 etc, – the number indicated hardness– can work from about –70 to about 310°C. Silastic NPC-40 and 80 do not require post cure and can be hot air cured. TR grades have better tear strength.

They are based on MVQ.

Among the high performance grades, 35U, 55U etc, have higher tensile strength. LT grade has better low temperature properties – brittle point –116°C (based on MPVQ). 2376U has better compression set resistance. 2351U has better flame retardancy – all these are based on MVQ.

Among special purpose grades, those for use in food processing areas may be Silastic 2000U contains FDI permitted additives. 1125U and HS50 are low in extractables and may contain only FDI permitted additives. SPG-30 is for sponge sheets, rods seals etc. LCS series are for lower compression sets – for seals and gaskets. SS-70 for rotary shaft seals HGS-70 for high green strength for excellent processing – for radiator hoses.

Silastic 1603, 1604, 1625 and WC-70 are for cables – all based on MVQ. They may contain flame retardant additives.

Grades containing FVQ are used where oil resistance is required. The grades are LS-40, LS 53U, LS-63U, LS 70, LS-422,LS-2249U, LS-2311U. Of these, the last two can withstand upto 232°C while the others, upto 177°C – all can withstand down to –61°C in the lower side.

Among the liquid silicones, many are RTVs (room temperature vulcanisables) and some need high temperatures for cure. Most of them are used for making flexible moulds. The grades are 3112, 3120, E RTV, J RTV, L RTV. For potting and encapsulations, Sylgard 170A and B (elastomers), Sylgard 182, 184, 186 etc, (resins). 3112 and 3120 are also used sometimes. Grade 93-500 is a clear solvent free polymer.

Sealants are also available – for formed-in-place gaskets. They are also based on liquid rubbers like Silastic 734 RVT, 738 RVT – used in pumps, pipe joints etc, besides potting, coating, tiles, connectors etc. In construction, weather resistant sealants are important – they are also based on RTVs e.g. 790, 888 grades. The latter is used in highway joints. Fire resistant barriers, roof and tank insulations, sealants of bathroom fixtures, sealants for window frame, fuel tanks in cars etc, coatings on PCBs, are some of the uses of liquid silicone rubbers.

Further, silicones are widely used in medical areas – mainly in implants, prosthetics, medical adhesives, mould making for bio engineering components.

Poly urethanes:

Strictly speaking poly urethanes cannot be considered as high performance elastomers as their only high performance characteristic is their exceptional abrasion resistance and other mechanical properties which cannot be matched by any other rubber. They are often considered along with silicones mainly due to the possibility of processing them in liquid state, like the latter while offering excellent properties achievable with solid rubbers – this is not possible with other elastomers. They may also offer solvent resistance and ozone resistance. Upper service temperature range does not exceed 120°C.

Like silicones they are also produced by step polymerization. Poly urethane formation is a step polymerization but not condensation polymerization as no side products are formed during the polymerization – these polymerizations are often called poly addition reactions.

Poly urethanes can be processed (i) in liquid state (castable PUs), or (ii) in solid state like a conventional rubber (millable PU elastomers) or (iii) like a thermoplastic (TPEs based on PUs).

The reactions leading to the formation such polymers are fairly simple ones – the raw materials needed are: polyols, isocyanates and chain extenders.

Organic isocyanates are highly reactive substances – here the carbon in the NCO group is connected to two highly electronegative elements i.e. N and O through double bonds – hence it is highly short of electrons and hence is prone to attack by nucleophiles - even weak nucleophiles can attack it – e.g. water.

$$-N{=}C{=}O + H{-}OH \longrightarrow -NH{-}\underset{\underset{OH}{|}}{C}{=}O \longrightarrow -NH_2 + CO_2$$

If H in water is replaced by R (i.e. alcohol), the product will be $-NH{-}\underset{\underset{OR}{|}}{C}{=}O$ If the reactants are glycol (instead of alcohol) and diisocyanate, a linear polymer will result.

This link is called urethane link. The urethane link has one more active hydrogen (i.e. the one attached to the N atom) – this can be attacked by another isocyanate group and this gives a cross link called allophanate link.

Similarly, if instead of glycol a diamine is used, the links formed will be urea link (linear polymer) –NH CO NH– and this will have two active hydrogen groups-these can again be attacked by isocyanates to give biuret cross link.

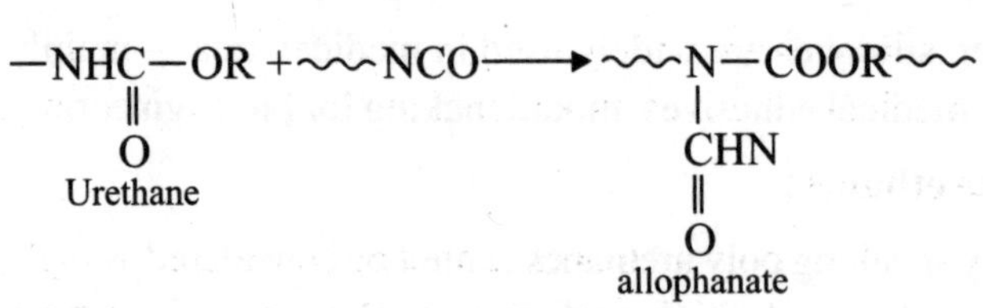

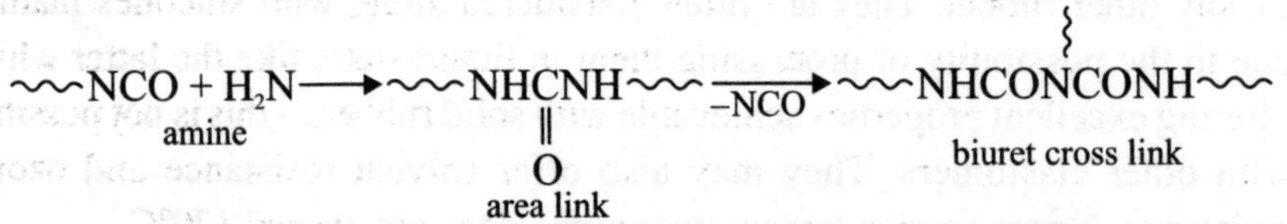

$$\underset{\underset{\text{Urethane}}{\underset{O}{\|}}}{-NHC}{-}OR + \sim\!\!\!\sim NCO \longrightarrow \sim\!\!\!\sim\underset{\underset{\underset{\text{allophanate}}{O}}{\underset{\|}{CHN}}}{N}{-}COOR\sim\!\!\!\sim$$

$$\sim\!\!\!\sim NCO + H_2N \longrightarrow \sim\!\!\!\sim\underset{\underset{\underset{\text{area link}}{O}}{\|}}{NHCNH}\sim\!\!\!\sim \xrightarrow{-NCO} \sim\!\!\!\sim\underset{\text{biuret cross link}}{NHCONCONH}\sim\!\!\!\sim$$

Simple reactions between glycol/diamine (chain extenders) and diisocyanate gives a product with many urethane/urea links – in these, hydrogen bonding will be extensive (like in nylons) and this will make the polymer rigid. Hence the hydrogen bonds must be separated from each other as much as possible to give elastomers – for this purpose we use polyols – which are low molecular weight polymers containing poly esters or poly ethers. Polyols are also available from natural sources like some vegetable oils– they do not give high strength PUs and hence is restricted to coatings mainly.

Reaction between polyols and isocyanate gives soft segments in the polymer (this gives the elastic properties) while isocyanate with chain extender gives hard segments (hardness comes from hydrogen bonds) – a combination of hard and soft segments give an elastomeric network.

Hardness of the resulting polymers can be adjusted by varying the type and amount of polyols and isocyanates and chain extenders – this is very much unlike other rubbers as in these cases filler content determines hardness mainly. The hard segments play the role of reinforcing fillers in poly urethanes. Further, strength of a PU comes from hydrogen bonding in the hard segments and not from cross linking which is the case with other rubbers. Further, cross linking does not need external agents as reaction between urethane and urea link with isocyanate itself will give cross linking. Thus compounding of these polymers will be required only for cost reduction and not for improving the hardness or elastic properties.

Commonly used diisocyanates are toluene 2,4 or 2,6 diisocyanate (TDI) or naphthalene 1,5 diisocyanate (NDI) or 4,4' diphenyl methane diisocyanate (MDI). They can be attacked easily by water or other nucleophiles and hence must be rigorously protected from water and other such reagents otherwise toxic vapour emissions will result.

Polyols are often low molecular weight polyesters like poly ethylene adipate with molecular weight of about a few thousands or poly butylene glycol or poly propylene glycol or hydroxyl terminated poly butadiene.

Chain extenders are glycols (e.g. butane diol) or diamines like MOCA – 4,4' methylene bis (orthochloroaniline) also called, bis (3 chloro 4 aminophenyl) methane.

Castable PUs:

Casting of PUs can be easily done by reacting the polyols, isocyanate and chain extender. A common variation is through the formation of a pre polymer obtained by reacting polyol and diisocyanate. This is then chain extended at the moulder's end. Often the supplier gives the pre polymer and the chain extender as per the requirement of the moulder which are mixed and cured by the moulder.

The castable PUs may be of the following types:

(*i*) **Unstable pre polymer type:** Here the pre polymer contains free iso cyanate groups which are highly reactive – they are mixed with chain extender and poured into moulds where it is cured at room temperature for one day or in presence of catalyst (tin octanoate generally) and at high temperatures in about an hour or more. This may be followed by post curing.

(ii) **Stable pre polymer type:** Here the isocyanate in the pre polymer will be end capped and this after mixing the chain extender and on heating will liberate the free isocyanate which then reacts to give the product-the advantage ill be better safety.

(iii) **One shot system:** In this system the three components are mixed in the mould and cured – the disadvantages are evolution of NCO vapours and the difficulty in controlling the structure of the resulting product – this is used in rare cases – the main advantage being low viscosity raw materials which enable low temperature processing.

(iv) **One pot system:** Here the prepolymer and isocyanate are mixed before transporting, at the PU producer end itself and the end groups are capped – on pouring into the mould and adding catalyst like acid the isocyanate becomes free and they react in the mould to give the product. In a way this is similar to stable prepolymer type – the only difference is the chain extender is also mixed at the PU producer end itself. The main advantages of this system is the low possibility of NCO vapour emission.

Diamines give higher hardness and strength while glycols give better flexibility. Poly ether polyols (except for those based on poly caparolactone, give better flexibility and poorer chemical resistance (as ester links in the main chain can be hydrolysed) while poly ether polyols give better chemical resistance.

Effect of the Stoichiometry on the nature of the PU formed:

Stoichiometry of the PU forming reaction is important in deciding the properties of the PU formed. For this we must consider the ratio of the number of isocyanate groups and active hydrogens.

If the [NCO]/ [OH] (from the polyol) ratio is 1:1 the resulting prepolymer cannot have excess isocyanate groups – the prepolymer cannot accommodate chain extender. Hence, the molecular weight may increase and the polymer will be linear and will have low hardness as the urethane links will be widely spaced.

If the ratio of [NCO]/[OH] (from polyol) is >2:1, more chain extender can be accommodated and this also will increase the concentration of the urethane links leading to higher hardness elastomers.

If the ratio of [NCO]/[active H] (here, H is from polyol and chain extender put together) is close to 1, at a slight excess of H, a thermoplastic rubber or millable rubbers will result. If the isocyanate content is slightly more, the excess of isocyanate will react with urethane link or urea link and produce cross linked polymer.

The one shot process can be used for RIM (reaction injection moulding) products, foams, etc.

Cross linking: Cross linking may be achieved by one of the following methods:

(*i*) by use of a tri functional chain extender like glycerol – this can give a linear polymer which can react with excess isocyanate and give urethane cross links which are not so thermally stable – this system may be used for making some foam products.

(*ii*) by use of excess isocyanate – this can lead to reaction between urethane or urea link with the excess isocyanate and produce allophanate or biuret links respectively.

(*iii*) by use of peroxides – they can cross link the millable gums in a way similar to cure of EPDM etc, and these cross links are stable to heat unlike allophanate or urethane cross links. Double bonds can be built into the systems by using allyl glycidyl ether or tri methylol propane, mono allyl ether etc, as chain extenders.

The resulting polymer will have double bonds which can be cured by sulphur and accelerators in a way similar to a diene rubber.

Properties: These mechanical properties of these rubbers can be adjusted by changing the nature of the polyol, chain extender etc, and changing the stoichiometry. Fillers may be used only for extending and not for improving the properties.

The strength comes mainly from hydrogen bonds. More concentration of H bonds will lead to greater hardness. Regular cross link distribution leads to excellent tear strength. Modulus can be changed through a wide range (3 to 600 MPa) and hardness can vary from Shore A 50 to Shore D 70 units. Throughout this hardness range, elongation at break and resilience remain constant.

PUs resist aliphatic hydrocarbons and are slightly affected by aromatic hydro carbons. Polar solvents will affect this rubber very much. Resistance to ozone and weather are good but thermal stability is limited – some grades and not all, can withstand upto 110°C. Low temperature flexibility can be shown down to –70°C in some grades. Specific gravity can vary from 1.1 to 1.2.

Often in PUs increasing cross link density will not lead to increase in mechanical properties. This is because cross linking leads to disruption in H bond concentration – it must be remembered that it is H bonding that gives strength to these rubbers.

If water is used as a chain extender, it will react with NCO groups giving amine end group which will react with more isocyanate to give urea links – they give stronger rubbers.

The structure of these polymers consists of alternating hard and soft segments which are inter connected and behave in a way similar to block copolymers which are used as thermoplastic elastomers. In the hard segments, the closeness of H bonds lead to formation of virtual cross links. The two Tgs of the hard and soft segments can be observed in thermal characterisation methods like DSC. X ray studies show hard segments (is segments with some ordering) dispersed in a continuous phase of soft segments – the hard segments seem to have smaller sizes and hence may not be qualified to be called crystalline segments – the word para crystalline may be used for this material. Annealing may help in the formation of more ordered structures.

As seen earlier, the hardness can vary over a wide range by change of the polyol and iso cyanates and their stoichiometry. Reaction Injection Moulding can give polymers with strength approaching those of engineering plastics– it is not uncommon to see RIM with reinforcing fibres (RRIM) being used as replacements for metals in engineering components like car body parts, dash boards.

Compounding: Millable rubbers may to some extent, be reinforced by carbon black etc, in typical rubber processing equipment. Castable polymers may be reinforced by glass fibres etc, when processed by RIM. Otherwise castable and Thermoplastic poly urethanes do not need much of compounding. Castable PUs need post curing to optimize mechanical properties.

RIM: Reaction Injection Moulding (RIM) is a very unique processing technique developed exclusively for PUs. Though other polymer systems too can be processed by this technique they cannot be done at such high speeds seen in PUs- a RIM machine for PU cannot be used for other polymer systems as reacting systems, mixing characteristics etc, are different for each system. Other polymers which can be processed by this method are epoxy resin, unsaturated poly ester resin, some poly amides based on caprolactam type raw materials. PU RIM consists of two/three stream of liquids (either pre polymer + chain extender or polyol + isocyanate + chain extender respectively) impinging into each other under high pressures leading to molecular level mixing, in a mixer head from which they are injected into a mould – thus polymerization and shaping both occur simultaneously inside the mould – thus the energy inputs needed for shaping get reduced very much and hence moulds may much lighter. Since poly urethane reactions may be speeded up very much by using catalysts like tin octoate etc, large products may be moulded in a short time.

Products may be small ones like shoe soles – they can be moulded in large number at a time – example even 25 pieces in about 10–15 minutes. Alternatively it can be a large product *e.g.* a car door – this can be moulded in a short time of about 5 minutes or less.

Variations of this process can be used for making foamed products – very large blocks of PU foams can be made in a very short time (5–10 minutes) with a very little labour input unlike latex foams which have a large lead time and many workers for production. Further very low densities which cannot be produced by latex route can be made through PUs. Thus in foams like mattresses or vehicle upholstery latex foams have been largely replaced by PU foams though the raw material cost may be 20 times more than the latex foam.

Foams of varying rigidities may be produced easily – from very soft cushions to fairly rigid foams used for packaging and thermal insulations can be made easily – just the reactants have to changed for varying the hardness.

Uses: PUs have extensive uses in foams, coatings and adhesives. A few mechanical goods are also made using PUs – hydraulic seals, shock absorbers, bearings, clutches, solid tyres, besides shoe soles, printers' rollers, rubber bands etc.

Compounding Ingredients

Rubbers cannot be used alone as they do not have the necessary properties expected out of them. There is, hardly any product where a rubber can be used alone – *e.g.* some adhesive formulations. Hence, they are almost invariably compounded.

The additives may be of the following broad classes:

(*i*) cure/vulcanizing systems – without them the elastic properties cannot be achieved.

(*ii*) reinforcements – they are needed for improving strength. Along with them processing aids are also added to make the rubber processible.

(iii) anti degradants – to protect the rubber against degradation by heat ageing, ozone, oxidation, weather.

These are further classified:

Cure systems: vulcanising agents, accelerators, activators.

Reinforcements: fillers, plasticizers, other processing aids.

Anti degradants: mainly anti oxidants and anti ozonants.

Apart from these, some other ingredients may also be included in some compounds.

Vulcanisation/Cure systems:

Vulcanisation of NR was discovered by an accident – sulphur fell on a molten NR and after some time it was noted that the portion where sulphur fell had much greater strength and better elastic properties. This was called vulcanization as Vulcan is the god of fire in Greek mythology – strengthening by treatment by heat.

During vulcanization, the polymer chains are cross linked and this leads to prevention of flow and renders any applied deformation elastic – this is also called curing. As the cross link density increases static modulus, dynamic modulus, hardness increase. Tensile strength, tear strength, fatigue life, toughness, all increase and attain a maximum and then fall with increase in cross link density. Hysteresis, permanent set and friction coefficient decrease with cross link density.

An ideal cure system should be of low cost, non toxic and be able to provide a reasonable fast cure without scorch danger.

For the diene rubbers, sulphur continues to be the most commonly used curative. Sulphur, available in nature is ground to the necessary particle size and used. Sulphur is mined at Sicily, Turkestan, USA, Poland besides a few other countries. Sulphur as available in nature, is in rhombic form. It is sparingly soluble in a few rubbers – *e.g.* in NR it can dissolve to about 1.6% at room temperature at 100°C the solubility may increase to 7%. The partial solubility of S in rubbers causes a problem of blooming of sulphur on the rubber if its content exceeds 0.5% . Blooming leads to spoilt appearance of products besides reducing tack in built up products.

Insoluble sulphur is obtained by polymerizing sulphur to give a molecular weight of a few lakhs – it is insoluble in solvents and in rubber and is processed at temperatures below 110°C. At higher temperatures it is converted to rhombic sulphur. Insoluble sulphur does not bloom and hence does not impair the tack property and it does not reduce scorch safety of compounds during storage.

Sulphur can be replaced by elements similar to it, i.e. selenium and tellurium – they improve heat resistance of the product.

Further, sulphur can be partially replaced by sulphur donors like dithio dimorpholine (also called bis morpholine di sulphide and TMTD etc, they give out free sulphur during cure, by decomposition on heating - advantages may be elimination of blooming and in products in contact with food.

Phenolic resins: Reactive phenolic resins can cure diene rubbers by cross linking in presence of the double bond in the rubber and then getting attached to the rubber molecular chains. They provide superior heat resistance – hence used for products requiring this property – *e.g.* butyl rubber based tyre curing bladder which has to withstand upto 165°C inside a tyre curing press. The curing requires activation – by metal halides *e.g.* $ZnCl_2$ This is not added directly, but as a small amount of CR (5 phr) in the compound which during cure reacts with ZnO (3 phr) and gives the $ZnCl_2$. Cure temperature for making the bladder is about 160–200°C.

Use of brominated phenolic resin eliminates the need for CR in the formulation – this is reactive on its own for this cure.

Epoxy resins can be used as curing agents for carboxyl terminated rubbers.

Quinone dioxime (QDO): They can also react in a way similar to phenolic resin with diene rubbers. The curing is scorchy and hence requires modification. The actual curative is para dinitroso benzene which is obtained by oxidation of quinine dioxime by PbO_2/Pb_3O_4. This also gives good heat resistance. QDO can also cure poly sulphide rubbers.

Mechanism of Sulphur Cure of NR without accelerator:

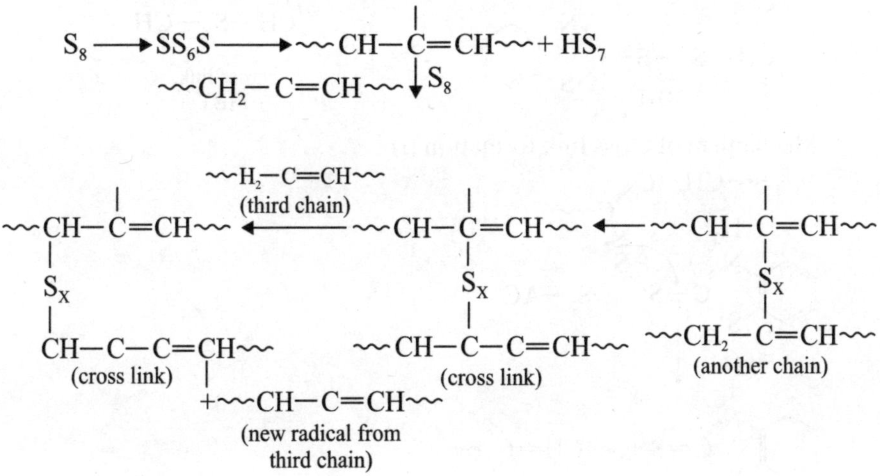

$$S_8 \longrightarrow SS_6S \longrightarrow \sim\sim CH-\overset{|}{C}=CH\sim\sim + HS_7$$
$$\sim\sim CH_2-C=CH\sim\sim \downarrow S_8$$

$$\sim\sim H_2-C=CH\sim\sim \text{ (third chain)}$$

$$\sim\sim CH-\overset{|}{C}=CH\sim\sim \longleftarrow \sim\sim CH-\overset{|}{C}=CH\sim\sim \longleftarrow \sim\sim CH-\overset{|}{C}=CH\sim\sim$$

$$\underset{\underset{\text{(cross link)}}{CH-C-C=CH\sim\sim}}{\overset{|}{S_X}} \quad \underset{\underset{\text{(cross link)}}{\sim\sim CH-C-C=CH\sim\sim}}{\overset{|}{S_X}} \quad \underset{\text{(another chain)}}{\overset{|}{S_X}}{\sim\sim CH_2-C=CH\sim\sim}$$

$$+\sim\sim CH-C=CH\sim\sim$$
(new radical from third chain)

Note :

1. Sulphur is available as S_8 naturally.

2. S_X because, during cross linking one or more sulphur atoms may be lost.

3. The reactions are thought to be preceiving though free radical formation though ionic mechanism (C^+) can also be considered.

Mechanism, with a thiazole accelerator:

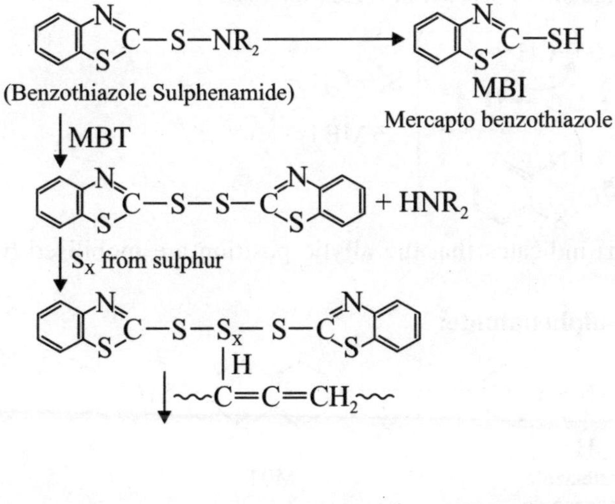

(Benzothiazole Sulphenamide) C—S—NR$_2$ $\longrightarrow$ C—SH **MBI**

Mercapto benzothiazole

$\downarrow$ MBT

C—S—S—C + HNR$_2$

$\downarrow S_X$ from sulphur

C—S—S$_X$—S—C

$\overset{|}{H}$

$\downarrow \sim\sim C=C=CH_2\sim\sim$

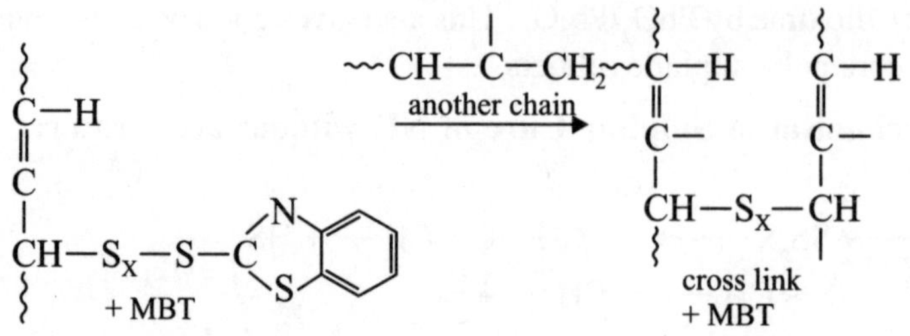

$$\text{Mechanism of cross link formation (i)}$$

Mechanism of cross link formation (i)

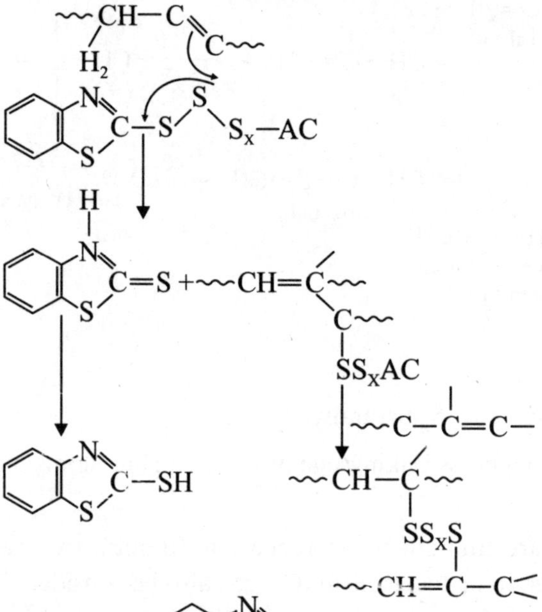

Note : AC =

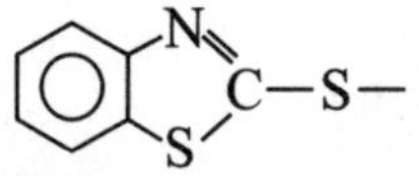

Another mechanism of cross link formation is (ii) :

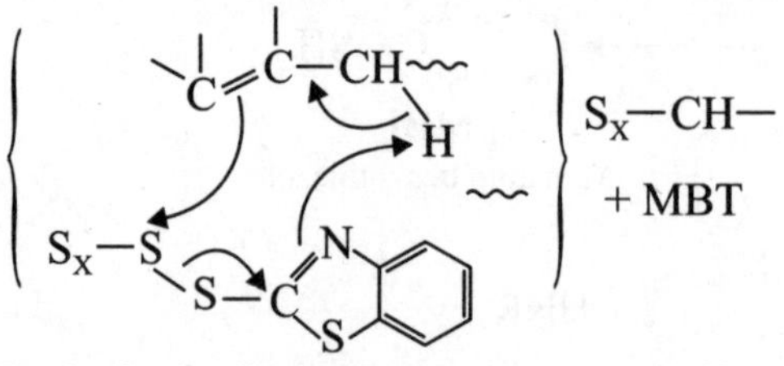

Both (i) & (ii) indicates that the allylic position are mobilised by the accelerator.

Action of a sulphenamide:

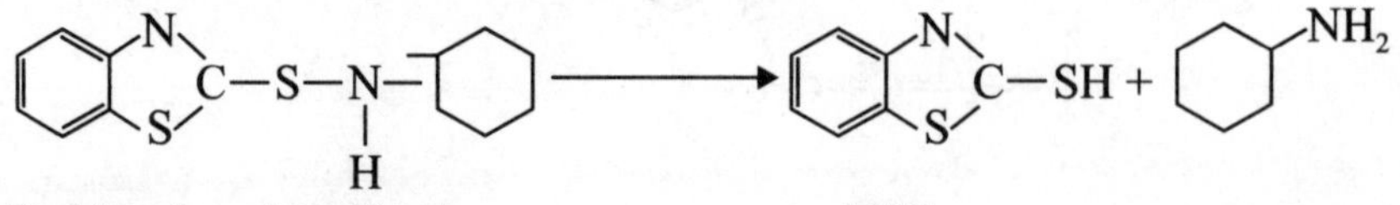

Cyclohexyl   benzthiazole
                  sulphenamde

MBT

MBT is the accelerator. As the heat is partialy used to decompose the sulphenamide, scarch safety is improved.

Further, pre-vulcanisation inhibition (PVI) can be useful in slowing down the pre-vulcanisation and provide. Scarch safety, (It reacts with MBT and puts it out of action.)

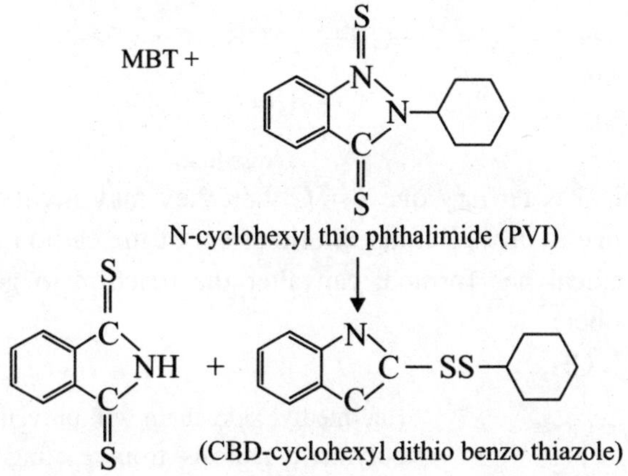

MBT +

N-cyclohexyl thio phthalimide (PVI)

NH +

(CBD-cyclohexyl dithio benzo thiazole)

Role of ZnO + stearic acid (activator):

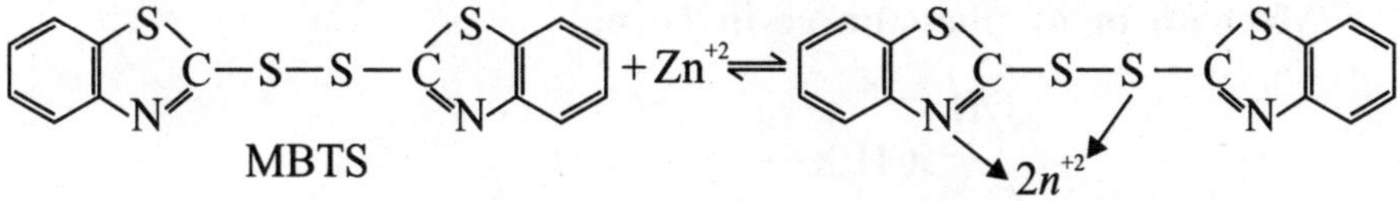

MBTS $+ Zn^{+2} \rightleftharpoons$

$2n^{+2}$

(Complexes the rubber with S
and accelerator & aids disperson
of curatives in rubber-thus,
improves cross linking efficiency

In Absence of activator	**In Presence of activator**

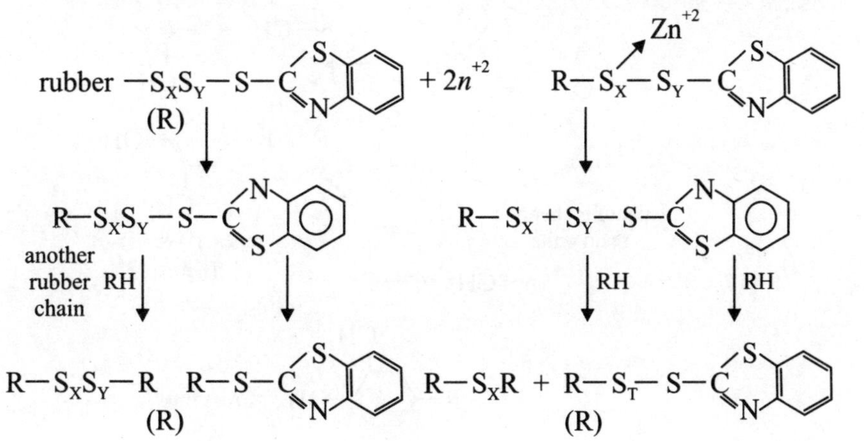

rubber $-S_X S_Y - S - C$ $+ 2n^{+2}$ (R)

Zn^{+2}

$R - S_X - S_Y - C$

$R - S_X S_Y - S - C$

another
rubber RH
chain

$R - S_X + S_Y - S - C$

RH   RH

$R - S_X S_Y - R$ $R - S - C$
(R)

$R - S_X R + R - S_T - S - C$
(R)

C–S bond is to be broken which
consumes a lot of energy hence,
this fragment does not help
in further crosslinking

as S–S bond is to be broken-which
consumes less energy hence can lead to
one more cross link formation thus,
better efficiency in use of curatives, here

Mechanism of peroxide curve:

$$P-O-O-P \longrightarrow 2PO$$

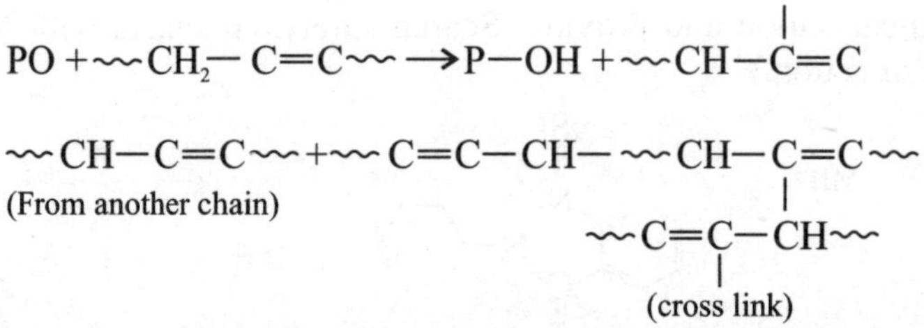

$$PO + \sim CH_2 - C=C\sim \longrightarrow P-OH + \sim CH - \overset{|}{C}=C$$

$$\sim CH-C=C\sim + \sim C=C-CH- \quad \sim CH-C=C\sim$$

(From another chain)

$$\sim C=C-CH\sim$$

(cross link)

If the peroxide radical is tertiary one $\sim$ C then they may involve in chain scission in preference to cross linking. Steric effect of the carbon next to the one where the radical has formed, can alter the reaction to go to scission (as in buthyl rubber)

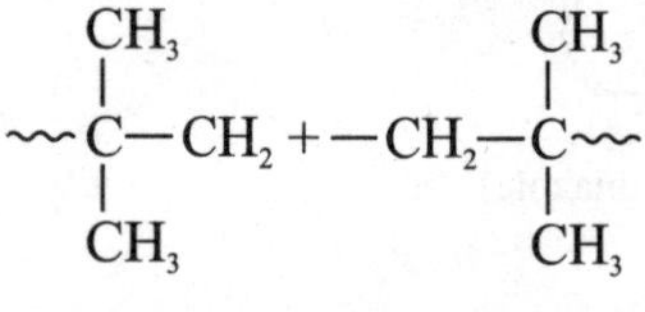

$$\sim \overset{\overset{\displaystyle CH_3}{|}}{\underset{\underset{\displaystyle CH_3}{|}}{C}} - CH_2 + - CH_2 - \overset{\overset{\displaystyle CH_3}{|}}{\underset{\underset{\displaystyle CH_3}{|}}{C}} \sim$$

(the methyl side chain will prevent the polymeric radicals from reacting with each other)

co-agent will destroy any tertiary radical which may form

Mechanism of phenolic resin cure:

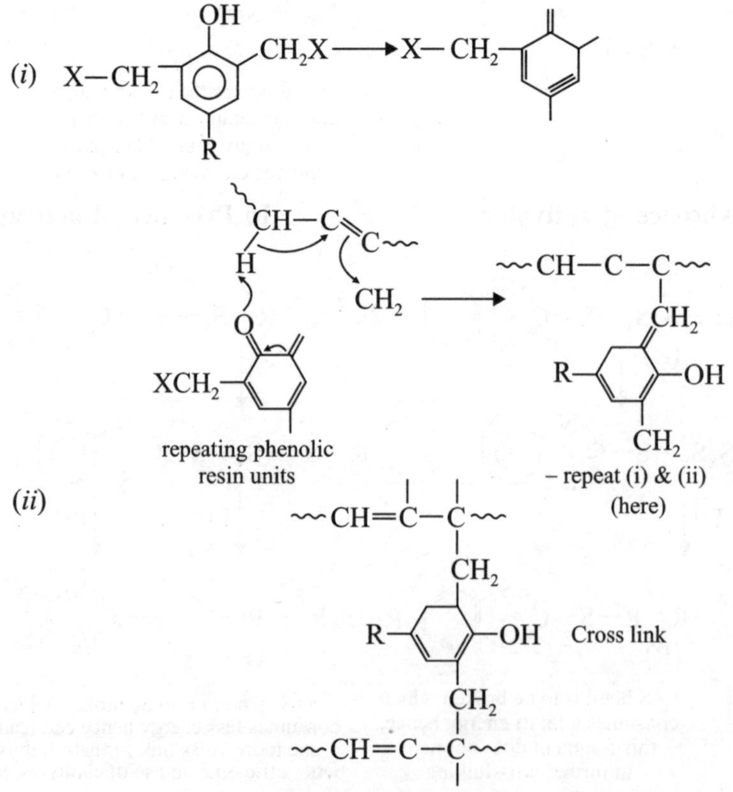

(i)

(ii)

repeating phenolic resin units

– repeat (i) & (ii) (here)

Cross link

Similar mechanism of occurs when resorcinol formaldehyde resin (generated in situ) reacts with rubber for promoting adhesion between rayon and nylon fibres with NR or SBR for tyre manufacture.

Oxime cure:

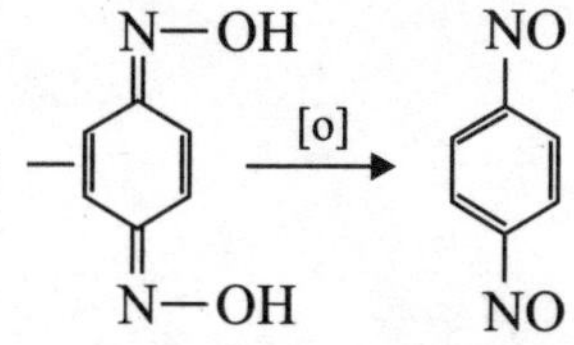

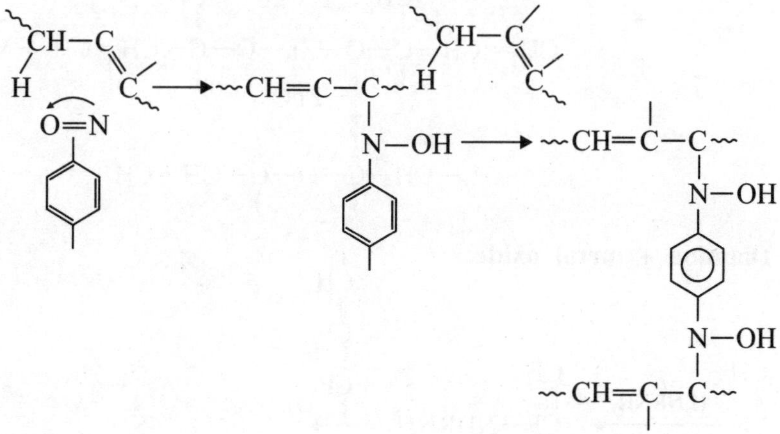

Bismaleimide cure: bis maleimide:

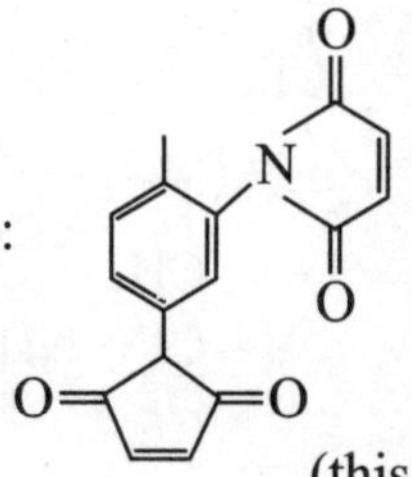

(this reaction requires, a peroxide initiator)

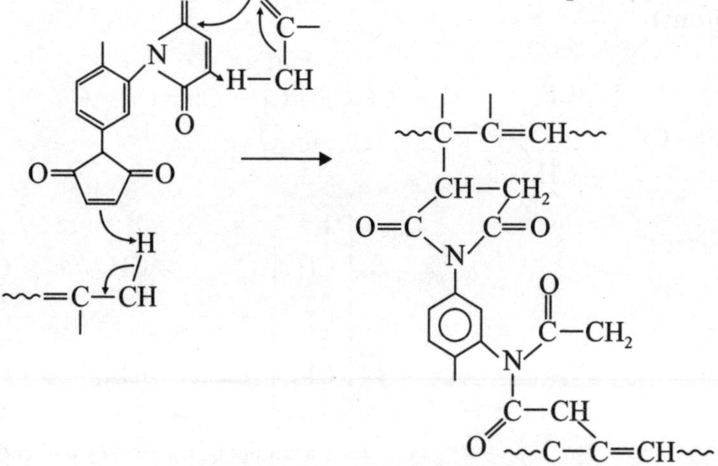

Metal oxide:

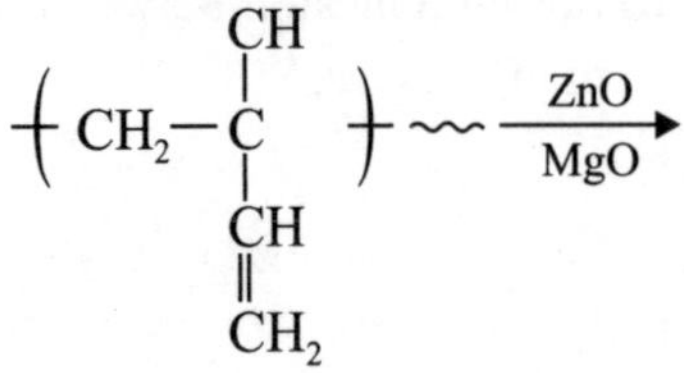

$$\left(CH_2-\underset{\overset{|}{\underset{\overset{|}{\underset{\|}{CH_2}}{CH}}}{\overset{|}{\underset{\|}{C}}{CH}}}\right)\sim\xrightarrow[\text{MgO}]{\text{ZnO}}$$

CR – its 1, 2, structure

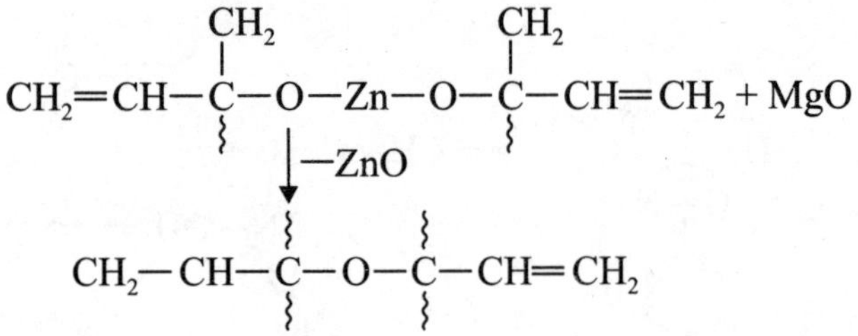

$$CH_2{=}CH-\underset{\overset{\}{\}}{\overset{CH_2}{\underset{|}{C}}}-O-Zn-O-\underset{\overset{\}{\}}{\overset{CH_2}{\underset{|}{C}}}-CH{=}CH_2 + MgO$$

$$\downarrow -ZnO$$

$$CH_2-CH-\underset{\overset{\}{\}}{C}-O-\underset{\overset{\}{\}}{C}-CH{=}CH_2$$

Diamine + metal oxide:

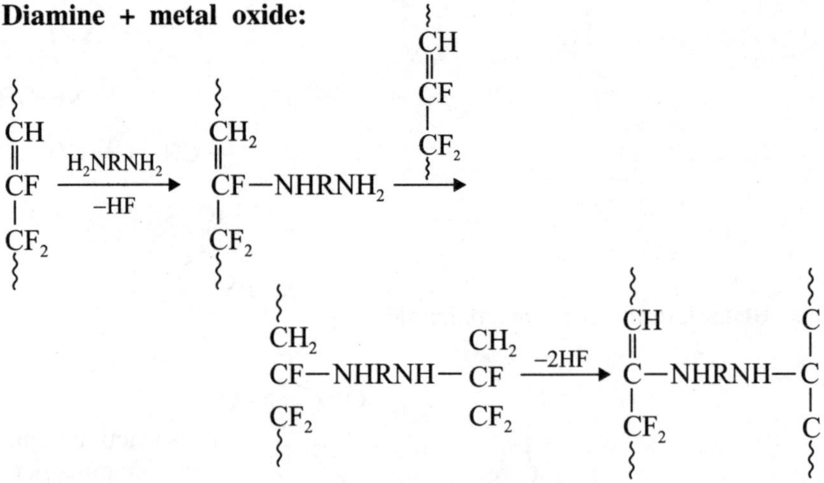

$$\underset{\overset{\}{\}}{\underset{\overset{|}{\underset{\overset{|}{CF_2}}{CF}}}{\overset{\}{\overset{CH}{\|}}}}\xrightarrow[-HF]{H_2NRNH_2}\underset{\overset{\}{\}}{\underset{\overset{|}{\underset{\overset{|}{CF_2}}{CF-NHRNH_2}}}{\overset{\}{\overset{CH_2}{\|}}}}\longrightarrow$$

$$\overset{\}{\overset{CH}{\underset{\|}{\underset{CF}{\underset{|}{\underset{CF_2}{\}}}}}}}$$

$$\underset{\overset{CF_2}{\}}}{\underset{CF-NHRNH-}{\underset{\overset{CH_2}{}}{}}}\,\underset{\overset{CF_2}{\}}}{\underset{CF}{\overset{CH_2}{}}}\xrightarrow{-2HF}\underset{\overset{CF_2}{\}}}{\underset{C-NHRNH-}{\overset{\overset{CH}{\|}}{}}}\underset{\overset{C}{\}}}{\underset{C}{\overset{\}}{}}$$

(in vitons)

Further,

$$\sim CF_2CF{=}CH-CF{=}CF\sim$$
$$+$$
$$\sim CF{=}CH\sim$$

$$\sim CF_2-CF\underset{CF-CH}{\overset{CF=CF}{<}}\!\!>CF \longleftarrow$$

$$2\left[\underset{\overset{\}{\}}{\underset{\overset{|}{\underset{\overset{|}{CF_2}}{CF}}}{\overset{\}{\overset{CH}{\|}}}}\right]\underset{-H_2O}{\overset{+H_2O}{\rightleftarrows}}\underset{\overset{CF_2}{\}}}{\underset{C{=}NRN{=}C}{\overset{\overset{CH_2}{|}}{}}}\underset{\overset{CF_2}{\}}}{\overset{\overset{CH_2}{|}}{}}$$

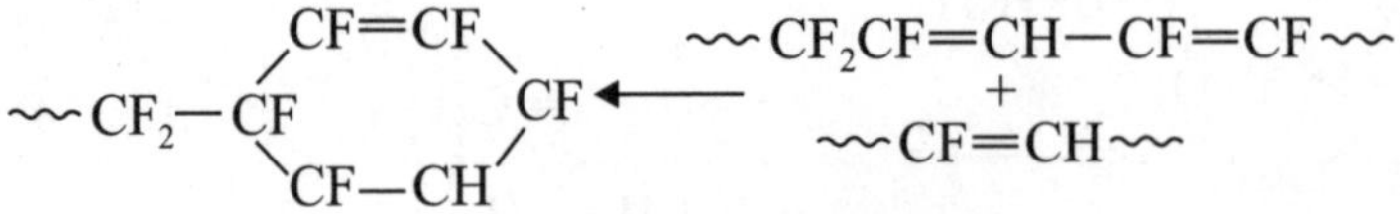

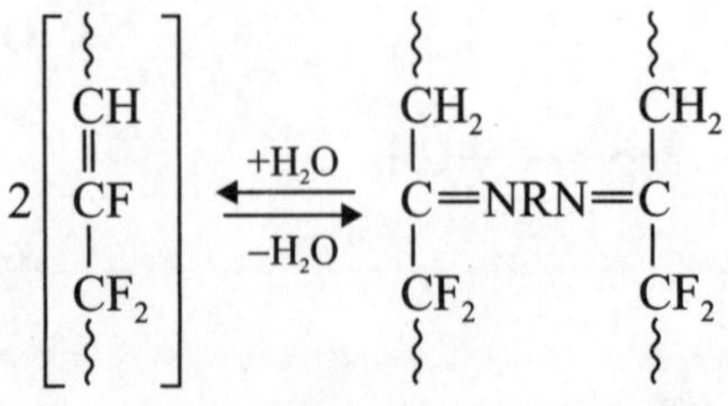

($\Rightarrow$ water must be removed by post curing)

Urethane cure of diene rubbers:

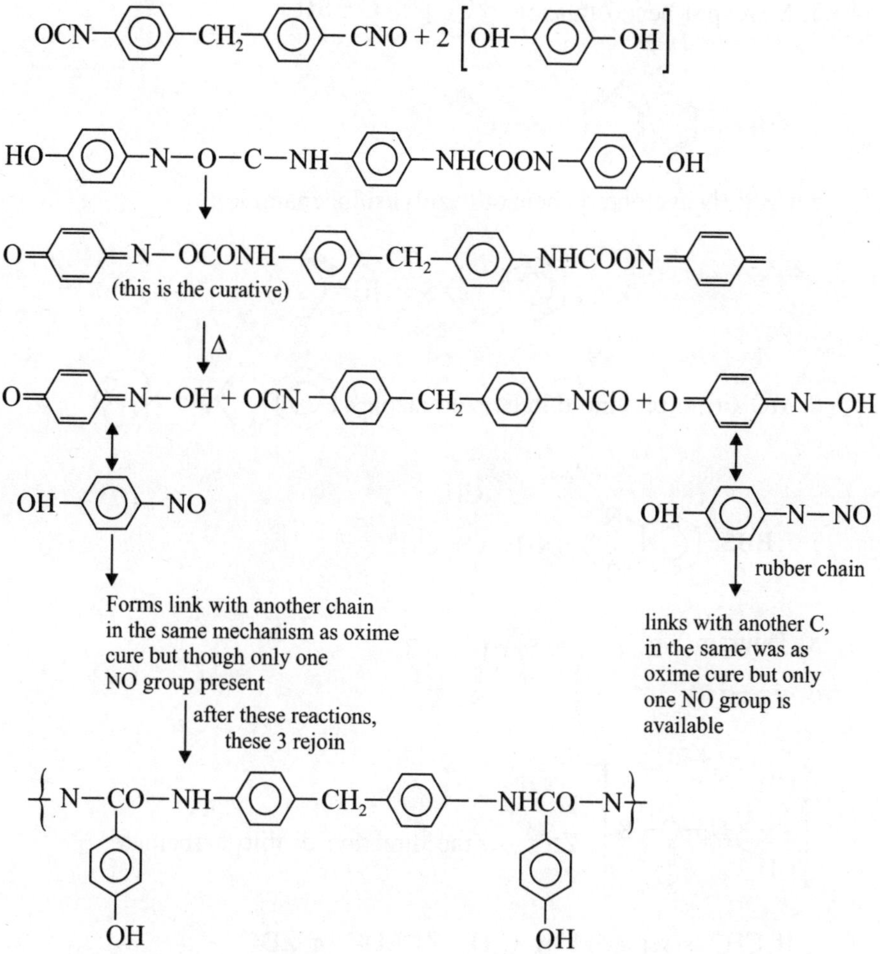

$$OCN-\!\!\bigcirc\!\!-CH_2-\!\!\bigcirc\!\!-CNO + 2\left[OH-\!\!\bigcirc\!\!-OH\right]$$

$$HO-\!\!\bigcirc\!\!-N-O-C-NH-\!\!\bigcirc\!\!-NHCOON-\!\!\bigcirc\!\!-OH$$

$$O=\!\!\bigcirc\!\!=N-OCONH-\!\!\bigcirc\!\!-CH_2-\!\!\bigcirc\!\!-NHCOON=\!\!\bigcirc\!\!=$$
(this is the curative)

↓ Δ

$$O=\!\!\bigcirc\!\!=N-OH + OCN-\!\!\bigcirc\!\!-CH_2-\!\!\bigcirc\!\!-NCO + O=\!\!\bigcirc\!\!=N-OH$$

$$OH-\!\!\bigcirc\!\!-NO \qquad\qquad OH-\!\!\bigcirc\!\!-N-NO$$

rubber chain

Forms link with another chain
in the same mechanism as oxime
cure but though only one
NO group present

links with another C,
in the same was as
oxime cure but only
one NO group is
available

after these reactions,
these 3 rejoin

$$\left\{\!N-CO-NH-\!\!\bigcirc\!\!-CH_2-\!\!\bigcirc\!\!-\!-NHCO-N\!\right\}$$

OH OH

Structure of accelerators:

1. Hexamethylene tetramine:

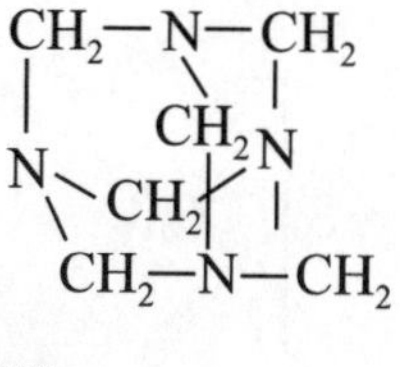

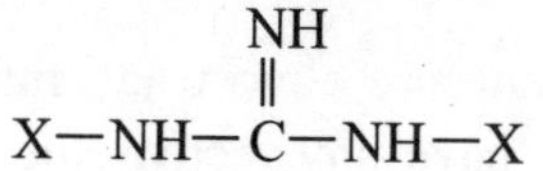

$$\begin{array}{c} CH_2-N-CH_2 \\ | \quad CH_2N \\ N \diagdown CH_2 \diagup | \\ CH_2-N-CH_2 \end{array}$$

2. Guanidine:

$$X-NH-\overset{\overset{\displaystyle NH}{\|}}{C}-NH-X$$

If X is Ph—DPG

If X is $\bigcirc\!\!-CH_3$ —DOTG

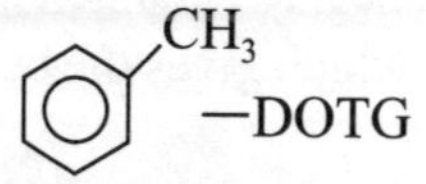

3. Mercapto benzo thiazole:

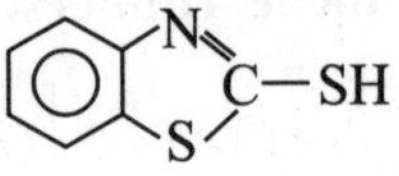

4. MBTS:

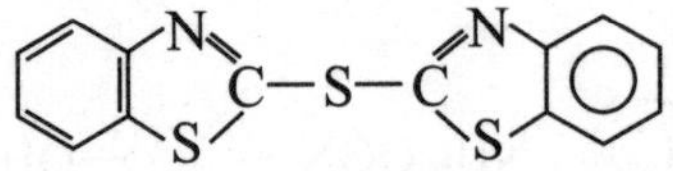

5. CBS: (N cyclohexyl benzothiazolyl/sulphenamide):

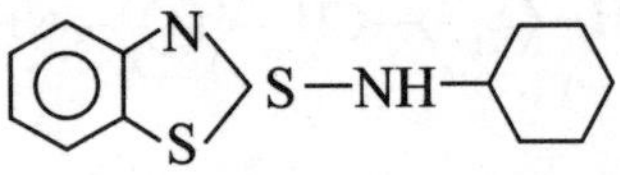

6. MOR (morpholino thio benzothiazole):

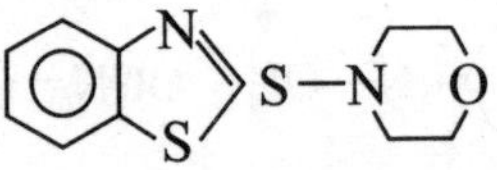

7. TBBS:

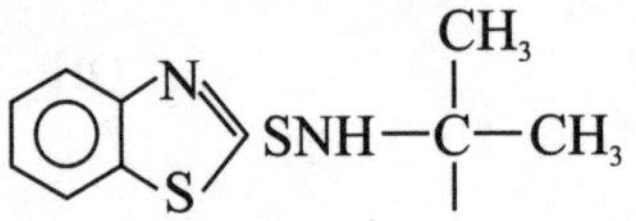

8. Thiurams: $R_2N-\underset{\underset{S}{\|}}{C}-S_x-\underset{\underset{S}{\|}}{C}-NR_2$

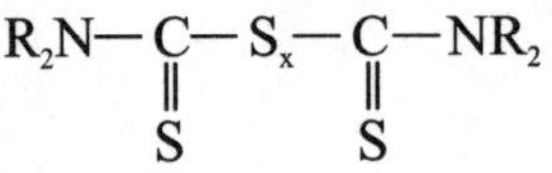

9. ZDMDC:

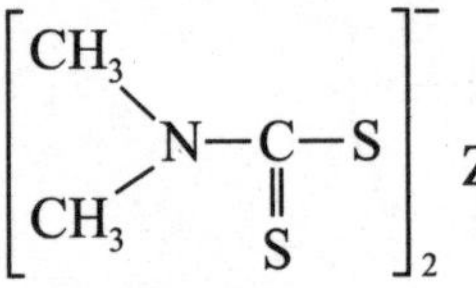

Zn^{+2} – Zinc dimethyl di thio carbamate

If CH_3 is replaced by C_2H_5, ZDEDC or ZDC

If CH_3 is replaced by C_4H_9, ZDBDC

10. **Xanthates:** ZIX

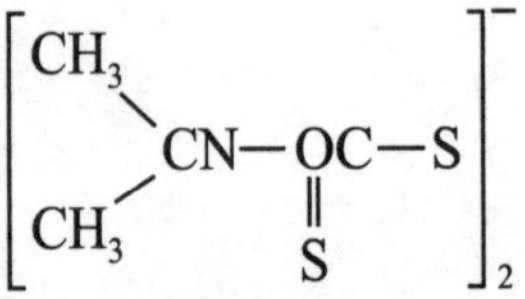

Zn^{+2}

Peroxides: Peroxides can cure almost any rubber – they are needed where the polymer does not have any rective site for curing like double bonds etc, – thus EPDM, silicones etc, can be cured by peroxide. They cross link by forming C–C cross link which gives the best compression set resistance and heat resistance.

Though IIR has very little unsaturation, still this cannot be cured by peroxide as in this case, peroxide causes degradation of the polymer.

Metal Oxides: ZnO can cure halogen containing rubbers like CR, ECO, CSM. ZnO removes the chlorine from these polymers and replaces with O atom from the oxide. Often the byproduct $ZnCl_2$ causes scorching of the rubber – hence it is neutralized by MgO. MgO however it not reactive enough to cure these rubbers effectively

Diamines: Hexa methylene diamine, piperidine etc, can cure rubbers containing Cl or F atoms. FKMs can be cured by hexamethylene diamine (its salt form-di carbamate). In FKM, HF is evolved as by product which is removed using MgO acid acceptor. Otherwise, HF will be harmful to the rubber.

Diisocyanates: They are used for urethanes only. Another isocyanate based cure system has been developed recently – this is an adduct of QDO and diisocyanate – this on heating decomposes to give isocyanate and 2 molecules of QDO which can cure the diene rubber – this system (one brand name is Novor) is useful for curing thick products which are prone to uneven cure by the curing methods used usually.

Other curatives may be sulphur + soap system used in some poly acrylate rubbers – the mechanism of this curing is totally different from S + accelerator of diene rubbers. Sulphur monochloride was also used for cold vulcanization of NR – the properties obtained were poor and hence it is not used anymore.

$Mg(OH)_2$, $Ca(OH)_2$ can cure rubbers like CSM, carboxylated NBR and give ionic cross links which are heat – fugitive and can re-form on cooling. Metal oxide cure of CSM can proceed at room temperature and this can be exploited in curing tank liners etc.

Sulphur vulcanization/curing:

Sulphur was the first cure system discovered – initially it was found that for attaining a satisfactory level of cure, 8 parts of sulphur for 100 parts of the rubber was needed and the process took 5 hrs at 140°C. Addition of zinc oxide decreased it to 3 hrs. Addition of organic accelerators decreased and the amount of sulphur used for cure and the cure time, to a few minutes. Thus, today, unaccelerated sulphur cure is not of any use. The mechanism of unaccelerated sulphur cure may be postulated as through free radical reaction.

Accelerated sulphur cure occurs through mobilization of the allylic positions and hence make them more active for the cure thereby speeding up the reaction. The mechanisms may be compared as follows:

Thus, accelerators lead to more efficient use of sulphur. Further, they help in curing faster which means that the rubber is exposed to lesser amounts of heat – this reduces the chances of deterioration of properties. Hence accelerators not only speed up cure but also improve mechanical properties.

Also, the number of sulphur atoms in a cross link is lesser when cured with accelerators – hence efficiency of cure is improved.

One of the first accelerator was aniline – this is toxic and hence had to be replaced by other substances – adducts of aniline—e.g. with carbon disulphide. Next came diphenyl guanidine (DPG) – it is still used but too slow. In the 30's a major break through in the form of thiazoles were discovered. After this, a further improvement in the form of sulphenamides also came up. Later, the ultra accelerators were discovered.

Accelerators may be classified in a number of ways.

By the chemical groups: thiazoles, thiurams, guanidines, dithio carbamates, xanthates, aldehyde-amine condensation products etc.

By speed: Ultra fast, fast, medium fast and delayed action, slow etc.

Primary and secondary accelerators: primary accelerators can be used alone while secondary accelerators are combined with primary accelerators.

Aldehyde amine condensation products:

1. Hexamethylene tetramine (HMT): This is a very slow accelerator – this is used for speeding up thiazole cure or they themselves may be speeded up by thiurams. It is hygroscopic and if not dried, cannot be easily dispersed into rubbers. For CR it is a fast accelerator.

2. Butyraldehyde-aniline condensation product: It is a liquid, medium fast, gives high modulus, and a flat vulcanization curve – mainly used in compounds where reclaimed rubber is present and it is not slowed down by carbon black loading.

3. Diphenyl guanidine (DPG): It is a slow accelerator, active above 135°C. It gives high modulus, strength and dynamic properties. It is slowed down by white factice, carbon black and kaolin.

4. Di-o-tolyl guanidine (DOTG): It is similar to DPG but safer to process. It gives poor ageing resistance. It is often used as activator for thiazoles.

5. MBT (mercapto benzo thiazole): This is a very important accelerator – gives low modulus, but good dynamic properties, a flat cure curve and good ageing resistance. It is a fast accelerator and has a higher tendency to scorch. It can be speeded up by thiurams, guanidines, dithio carbamates, CaO, MgO, etc. Acidic substances retard the cure.

6. MBTS (bis mercapto benzo thiazole disulphide): It is medium fast, low modulus and acidic accelerator – has a higher processing temperature and hence provides scorch safety – does not get speeded up by MBT but can help in scorch reduction in MBT containing compounds.

7. ZMBT (zinc salt of MBT): It is also a low modulus, fast accelerator. It does not bloom and hence suitable for light coloured mixes. All these thiazoles impart a bitter taste to the vulcanisate.

8. Sulphenamides (of benzothiazoles): The H in the SH group of MBT is replaced by amines – e.g. cyclohexane.

They may be used as primary or secondary accelerators . At higher dosages they do not reduce scorch safety but at the same time speed up the cure. They may provide good ageing resistance. They may be accelerated by ZDC or TMTD. A number of sulphenamides are available – CBS (cyclohexyl benzo thiazolyl sulphenamide), TBBS (t-butyl benzothiazolyl sulphenamide), NOBS or MOR (morpholino thio benzothiazole).

Processing safety is of the following order: CBS < TBBS < NOBS.

Sulphenamide based on the hypothetical dialkyl dithio carbamic acid (i.e. thiocarbamoyl sulphenamide) gives excellent processing safety while at the same time very high speed – hence may be called delayed action ultra fast accelerator.

9. Thiurams: They are tetra alkyl thiuram mono/disulphides. They are very fast accelerators. They have low scorch times but high cure rates. They are rarely used alone but only in combination with other accelerators. The disulphides also act as sulphur donors and hence lesser sulphur can be used in the mixes. They also impart good heat resistance to the vulcanisates. Anti oxidants act synergistically with these accelerators for heat resistance. Blooming may be a possibility but this can be suppressed by MBI (mercapto benz imidazole) etc. Scorch danger maybe reduced by adding MBT. Sulphurless cures are also possible using TMTD. They are non toxic and hence can be used in products for medical applications.

TMTM gives low modulus, and low compression set. Cure temperatures of 130–145°C. TMTD can be used for cure temperatures of 140–160°C for sulphurless cures and 110–145°C for sulphur cures. Tetra ethyl thiuram disulphides (TETD) is easier to disperse in rubber mixes.

10. Zinc dibutyl dithio phosphate: It is a liquid and a fast accelerator – suitable for EPDM compounds. It also gives good processing safety.

11. Dithio carbamates: Zinc dialkyl dithio carbamates: (alkyl maybe methyl or ethyl or butyl if methyl ZDMDC and if ethyl ZDEDC etc): They give medium modulus and are very fast in action. They are suitable for latex formulations and for high initial cure levels as in hot air cures, steam cures etc. ZDEDC is slower than ZDMDC. It can also be used as a secondary accelerators for guanidines. When the alkyl groups are higher, melting points decrease and they become less polar and hence may be more suitable for less unsaturated elastomers like EPDM, especially in blends with SBR etc.

12. Xanthates: Zinc isopropyl xanthate (ZIX) or sodium isopropyl xanthate (SIX) – they can cure even at room temperatures. They also give good ageing resistance.

For many rubbers, cure systems other than sulphur are used – they may use different accelerators – for example for CR, ethylene thiourea is an accelerator. Similarly for peroxide cures, a coagent is required – which helps in speeding up cure and increasing cross linking efficiency – the common chemical used for this purpose is tri allyl cyanurate (TAC).

Activators: For sulphur cure, presence of zinc oxide + stearic acid is also essential for improving the efficiency of cross linking – the mechanism of their action is shown in page 97.

They do not speed up cure but only increase the efficiency of cross linking. They help in dispersing the sulphur and accelerator in the rubber due to their intermediate polarity between the rubber and the accelerator by forming zinc stearate in situ – a significant point to be noted is that zinc stearate is not so useful as activator when added to the rubber – only zinc stearate formed in situ, can be useful.

Co-agent: During peroxide cure of any polymer, if any tertiary radical forms they may tend to undergo reactions other than coupling (which cross links) especially decomposition, leading to chain scission – this can be prevented by decomposing the tertiary radical immediately as it forms – this role will be fulfilled by co-agent – the main example for this is triallyl cyanurate (TAC). In the cure of butyl rubber by peroxide all the radicals which form will be tertiary – hence chain scission is more than radical coupling – hence butyl rubber cannot be cured by peroxide. In EPDM, some radicals which form will be tertiary – they can be neutralized by co-agent.

Fillers: Vulcanisation can improve the properties of a rubber but this is only to some extent. More improvements may be needed to make technically viable products – this requires reinforcements. Fillers also serve other purposes – they may act as extenders and reduce cost, improve processibility – especially, improve extrusion characteristics like reduced die swell or better surface finish, reduce swelling in oil etc, for seals, improve mechanical properties.

Fillers can be inert or active – strictly speaking no filler is purely inert or active. Active fillers are often reinforcing while inert fillers are mostly useful as extenders. Another classification can be made on the basis of black and on black fillers.

Carbon black: Carbon black is the most important filler for rubbers. The reinforcing ability of carbon black was discovered decades black. Today, a number of types of carbon blacks (about 50) are produced and each has its own uses – often, overlaps in properties can be seen.

Carbon black is made by either incomplete combustion of hydrocarbons or by thermal decomposition of the hydrocarbons. Carbon black production can be done by about three processes:

Furnace process: Most of the carbon blacks are produced by this method. In this, the partial combustion is done in horizontal furnaces. The raw material is liquid or gaseous aromatic hydrocarbons. A yellow flame is produced due to limited supply of air – this leads to partial combustion which causes soot to form – this is the carbon black which is present along with CO_2, N_2 H_2O vapour in the gas produced by the partial decomposition. This gas is cooled by water spray and passed through a series of cyclone separators from which the carbon black is collected. By increasing the air supply, the particle size and yield can be decreased. Yield can be 25-65% and particle size is 10-100nm. This process does not pollute the air excessively. The particle size is small and these particles have to be pelletised for easy handling.

Thermal Black: This forms about 5% of the total carbon black production. Unlike furnace black, this is formed by thermal decomposition of the hydrocarbons. Raw material is natural gas or generator gas or liquid hydrocarbons etc. The production is through two stages in a pair of chambers lined with refractory material. In the first chamber, decomposition occurs at about 1200°C (this temperature is achieved by complete burning of the fuel) – at this stage, the combustion is stopped and the second stage of decomposition starts where the production is completed. In the meantime the temperature of the second chamber is increased in a similar way. The effluent gas which takes away the carbon cools the first chamber and the second chamber now becomes heated and gets ready for the decomposition. When the first chamber again gets heated, the carbon which remained there will be burnt – thus the yield decreases – in fact, it is about 30–40% only. The particle size is larger – 120–500nm – this is difficult to pelletise.

Acetylene black: This is obtained by decomposition by heat. This is a very pure black with a high structure. This is rarely used.

Lamp black: This is obtained by particle combustion of liquid hydrocarbons in open pans – similar to the soot obtained by a lamp fuelled by burning paraffin. This black is not used much in the rubber industry. It is the oldest known carbon black and was the raw material for Indian ink.

Channel black: Natural gas is burnt in 'hot houses' containing a large number of burners, with a limited supply of air. The channels 'split' the flame and receives the black which deposits on it. The channels are moved to and fro and scrapers will remove the deposited black which is sent for further processing. The yield is very low. Some modified processes can help achieve higher yields – still, this forms a very small portion of the carbon black market. Channel black makes the rubber more electrically conductive.

The carbon blacks are fluffy and hence difficult to handle – they are pelletised to increase the density. The pelletisation is done by agitating the black with water and subsequent drying. Pelletised black is much easier to mix into a rubber than the fluffy black. The pelletisation causes many primary particles to agglomerate and this gives rise to particle structure. The spherical particles are fused together at the nodes and arrange into some irregular structure which leads to formation of empty spaces which are active sites for the rubber to be adsorbed. In addition porosity will occur in the particles. Further, the carbon black particles also have reactive groups like carbonyls, hydroxyl etc. due to surface oxidation during manufacture – they can also react with the rubbers. All these, together lead to a very good reinforcing action by the carbon black onto the rubber.

Important characteristics of the carbon black are particle size, particle structure, void volume, porosity etc. They can divided into 3 classes – extensive (those that depend on contact area available for the rubber – *e.g.* particle size), geometrical (structure) and intensive (interaction and reinforcing activity of the surface) factors. Particle size is very important in determining reinforcing ability – smaller size means more surface area available for interacting with the rubber and hence better reinforcement. The ASTM classification depends on particle size and structure – the first digit corresponds to size – in this, the highest number is 900-999 – this is the Medium Thermal black as per old classification – particle size is 200-500 nm.

800-899 is the Fast thermal black, with particle size in the range of 100-200nm. 700-799 is Semi Reinforcing Furnace black (SRF) – particle size 61-100 nm. 600-699 is the General Processing Furnace black (GPF) with particle size 49-60 nm. 500-599 is Fast Extrusion Furnace (FEF – particle size 40-48nm), 400-499 – FF black-31-39nm, 300-399 – HAF-high abrasion furnace – 26-30nnm, 200-299 is ISAF-Intermediate Super Abrasion Furnace – 20-25nm, and finally 100-199 is SAF-super abrasion furnace 11-19nm.

Particle size can be measured by BET method and Iodine Adsorption number.

High structure means more nodules per aggregate (often taken as particle) – this can be found by Di Butyl Phthalate (DBP) absorption number – this too can be done using a burette from which DBP is added to the black till all the nodules are filled by the liquid so that the mixture becomes a hard dough to which no more liquid can be mixed – this can be expressed as cc of DBP per 100 gm of the carbon black. Each nodule is a paracrystalline domain consisting of concentrically arranged graphite-like layer planes made of carbon atoms. It is possible to get same particle structure for two types of carbon black – the ability of DBP to fill the spaces will depend on both the space between the particles and the structure of the individual aggregate.

The reinforcement of a rubber by carbon black is a chemisorption process – this occurs at cure temperatures – adsorption of the rubber at the nodules – this is facilitated by the reactive groups at the surface. The milling of the rubber will lead to polymer chains getting cut and forming free radicals which easily graft themselves on to the carbon black particle surface. Depending on the nature of the functional group formed, the surface can be acidic or alkaline or neutral. Those which have acidic structure will retard cure while those with alkaline surface will accelerate cure.

On heating the black to about 2000°C the reactive groups will be removed – the reinforcing ability becomes very low.

The adsorption leads to formation of "bound" rubber – more the bound rubber, the more is the reinforcing ability. A rubber mix when treated with a good solvent for the rubber will form a gel – from this it will be difficult to remove all the rubber bound by the carbon black. Since chemisorption is accompanied by evolution of heat during the adsorption, we can expect the mixing of carbon black with rubber to raise the temperature - this is observed to be so.

In the initial stage of mixing, the rubber penetrates the nodules in the carbon black aggregate. If enough interactions are set up, the bound rubber will cement the aggregates further – this makes subsequent dispersion difficult. Thus, low structure – high surface area black is difficult to disperse but they are easy to incorporate. High structure black is difficult (slow) to incorporate but easy to disperse.

Bound rubber formation leads to loss of mobility of the rubber molecular chains – Tg increases slightly. Actually more rubber gets immobilised than what would be expected on the basis of bound rubber alone. A better term is "occluded" rubber. This is the rubber which finds itself in the internal void space of the structural aggregates. It is shielded from deformation of the rubber when it is strained.

More structure in the filler particle aggregate, the more will be the viscosity and lower melt elasticity.

The viscosity of a suspension of spherical particles in a medium of viscosity η_0 is given by:

$$\eta = \eta_0 (1 + 2.5c + 14.1c^2)$$ where c is the volume fraction of the rubber.

The actual increase in viscosity of the mix caused by the reinforcing filler is much more what will be predicted by the above equation.

This can be explained in terms of occluded rubber. The occluded rubber will become part of the filler – more the particle structure more will be the contribution of occluded rubber to C. If occluded rubber is shielded from deformation, then the total mass of rubber capable of reversible deformation must be decreased by occlusion of part of the rubber. Thus lesser mill and extrusion shrinkage will result. The viscosity will be more shear rate dependent if structure increases. At very low shear rates, secondary filler agglomerations are set up, which is disrupted by increasing shear stresses.

Scorch time may decrease with lesser particle size and structure as both these lead to more heating up of the rubber during mixing thus speeding up the cross linking. However, if the surface of the particle is acidic, this may retard cure. Further, carbon black opens up the S_8 rings leading to formation of more H_2S which can speed up the cure.

The effect of carbon black content on the processing and the vulcanisate properties are given below:

Tensile strength and modulus will increase with carbon black loading upto a point beyond which they will decrease. Resilience and elongation at break will decrease with carbon black loading. To what extent hardness of a rubber will increase with carbon black loading, is exemplified by the following example: For HAF black , for NR, at 45 phr loading hardness reaches 60 Shore A units. For SBR, for reaching 60 Shore A units, 45 phr will be required. For OESBR, it is 48 phr, for BR it is 56 phr, for IR it is 52 phr and for EPDM it is 45 phr.

This is based on a thumb rule which is widely followed by technologists which indicates the increase in hardness for NR for every few phr of carbon black loading – e.g. basic hardness of NR is said to be 40 phr and for every 2 phr of carbon black (HAF added) hardness will increase by 1 unit approximately.

Mechanism of reinforcement:

In many rubbers, the cross links are distributed unevenly in the mass of the rubber. The cross links ae introduced at random at high temperature where the chains may not be in the maximum probable confirmations – thus some of the chains may be stressed even after the cure. Such a solid if subjected to a stress, will start failing at their weakest chains (there it is the shortest as they are the most strained ones). In other solids too, the flaws which are present due to the manufacturing processes are the starting points of failure – thus the strength achieved are 20-1000 times lower than the expected values.

Now, if the most strained chains are allowed to slip, their tensions will be relieved and they will not break pre-maturely. Instead, more chains will now carry load and thus strength increases.

This can be depicted below:

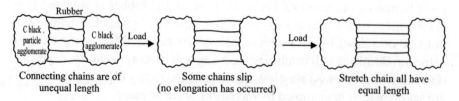

| Connecting chains are of unequal length | Some chains slip (no elongation has occurred) | Stretch chain all have equal length |

In this condition, if the chains are allowed to relax (if the stress is removed), they cannot return to their original positions – this leads to stress softening. Mullins effect – stressing for a second time needs lesser energy. Thus, the viscous component of the material behaviour comes to play and this leads to energy dissipation-leading to hysteresisence the high heat build up on carbon black filled rubbers in service and during mixing. Thus in carbon black reinforcement of a rubber, high modulus is due to the sites of high absorptive energy and high strength is mainly due to energy used in slippage-this can account for improved abrasion resistance.

The overall picture is that reinforcement is mainly due to introduction of very small sized flaws which will grow in size only when the loads are high. Till such loads are reached, the energy by stress will be used only for slippage and not for flaws to grow in size.

Non Black fillers: While carbon black reinforcement of a rubber is very unique in the sense that the reinforcing material is itself not a strong (mechanically) material – in other reinforcements as in FRP and in reinforced concrete, the reinforcing material is a strong material. Carbon black reinforcement is essential for rubbers but it comes with some disadvantages – it prevents colouring of rubber products (by other colours), this reinforcement always comes with reduction in resilience and also high heat generation during processing. Further carbon black itself is not a cheapening filler – its cost is not so low.

Non black fillers may solve some of these problems in a few cases though they can never substitute for C black completely. Further, many non black fillers are available in nature or by products of other industries and hence need not be produced at a high cost compared with carbon black.

Among non black fillers, silicas are reinforcing while the others are often non reinforcing or at best, semi reinforcing.

White fillers with particle size more than 5 microns are coarse ones – they do not improve mechanical properties – they may even deteriorate them – hence they are rarely used. Those between 1 and 5 microns do not reduce the properties that much – they may be called diluents or extenders e.g., whiting, limestone, soft clay. Those below 1 micron improve properties considerably – they are semi reinforcing ones – hard clay, precipitated $CaCO_3$. Those below 0.1 micron will improve properties very much – they are reinforcing e.g. precipitated and fumed silicas, fine types of calcium and aluminium silicates .

Besides these, we also have organic fillers – resins which can also improve mechanical properties.

As per ASTM D 1765 two letters and three digits are used to classify silicas and silicates – A or H indicates hydrated or anhydrous respectively, F or S or M medan fast or slow or medium rates of cure respectively and among the three digits the first one corresponds to particle size – others will be similar to carbon blacks.

Ground silica will have large particle size – they are got by grinding the naturally occurring silica. Hydrated or precipitated silica is obtained by hydrolyzing metal silicates with acids. Pyregenous or fumed silica is obtained by hydrolyzing $SiCl_4$ while in oxidizing flame at 1000-1800°C – in this method, moisture will be reduced considerably. There is a way by which another grade of silica similar to fumed silica is produced – by evaporating silicon dioxide in presence of carbon which gives pure silicon element which can be combined with oxygen to give silica – this is very pure. These methods also apply to other inorganic fillers.

Reinforcing ability depends on particle size, structure, porosity etc – similar to carbon black. In compounds where the total contact area less than 6 sq.m of the polymer-filler surface in 1 cm^3 of rubber, not much reinforcement will occur. At the same time, this contact cannot be increased by increasing the loading of coarse fillers. The coarser non black fillers can cause harm to the mechanical properties – this effect is less pronounced in carbon black. Particle structure is similar in concept used in carbon black in many ways. The crystalline structure of the aggregate determines the degree of anisotropy of the aggregate. More anisotropy leads to more viscosity, more hysteresis and more modulus in the compound. They are more difficult to incorporate into the rubber. The micro pores may absorb the accelerators etc, and retard cure – the rubber chains may not penetrate the pores of the inorganic fillers.

Moisture content in inorganic fillers can be considerable which may deteriorate electrical properties, reduce ageing resistance, formation of lumps during mixing, sticking to mill rolls etc.

Most of the inorganic fillers have no attraction to the organic rubbers. In silica fillers, silanol groups are present which interact with each other forming hydrogen bonds and hence they have no interaction with rubbers. Unreacted silanol groups are very few on silica surface. These groups can readily react with OH or halogens or amine groups. These are acidic and hence react with accelerators – this leads to retardation of cure. Silicone rubbers are almost invariably reinforced by silica and purity of the filler is important. This is also true for a few other speciality rubbers. Silicones are reinforced by fumed silica and the interactions between this filler and silicone rubber is stronger than those between organic rubbers and carbon black. This is manifest as hardening of the compounds during storage – structuring – this necessitates freshening of the compound for further processing.

Inorganic fillers must be surface treated for obtaining good reinforcement. The chemicals used are often high molecular weight organic substances containing polar end groups as in silane coupling agent for silica filler. They make the filler surface hydrophobic and hence improve compatibility with organic rubbers. The polar end will be attracted towards the filler surface. The interactions are improved at the vulcanization temperatures where some reactions, as shown below, will occur leading to good reinforcement. At the same time, abrasion resistance, fatigue resistance, and compression set will improve while heat build up decreases due to the coupling agent.

The active end group in the silane coupling agent can be vinyl or mercapto or amine or methacryloxy – mercapto ones are effective for sulphur cure while the others are for peroxide cure. The mercapto silane may cause scorch problems and this may necessitate the use of more delayed action accelerators in the formulations.

$$\begin{array}{l} \text{SiO}_2 \text{ Particle} \left\{ \begin{array}{l} \text{(H}_2\text{O)} \\ \equiv \text{Si—OH} + (\text{CH}_3\text{O})_3 \text{ Si(CH}_2)_3 \text{ SH} \xrightarrow{-\text{CH}_3\text{OH}} \\ \text{(H}_2\text{O)} \end{array} \right. \end{array}$$

$$\text{SiO}_2 \equiv \text{Si—OH} + \text{HO} + \underbrace{\text{HOSi(CH}_2)_3 \text{ SH}} \quad \Big\downarrow -\text{H}_2\text{O}$$

$$\text{SiO}_2 \equiv \text{Si—O—Si—(CH}_2) \text{ Si} \left\{ \xleftarrow[\substack{(\text{rubber}) \\ \text{during cure}}]{-\text{CH=CH—CH}_2\sim} \right.$$

$$\text{SiO}_2 \equiv \text{Si—O—Si—(CH}_2)_3 \text{ SH} \quad \Big\downarrow \text{HO}^-$$

Bis (tri ethoxy silyl propyl) tetra sulphide is another coupling agent used for sulphur cure. Vinyl silane is used for cable production, based on EPDM with peroxide curing.

Amino propyl tri ethoxy silane is good for sulphur cure and peroxide cure of diene rubbers. They can react with sulphur compounds and initiate the decomposition of peroxide. They are expensive and hence used only in poly urethane rubbers. For CR or CIIR, chloro propyl triethoxy silane is useful.

Silanes can easily hydrolyse and are liquids and hence may be added along with filler which are not treated by other substances. They must be stored in dry conditions. They should be added with fillers before other additives are added. Fillers pre treated by silanes are also commonly used. Similar to silanes titanates are also available for functioning as coupling agent in a way similar to silanes. Similarly polymeric coatings for fillers are also used – e.g. carboxyl terminated poly butadiene coated over $CaCO_3$ can give as good a reinforcement for diene rubbers as a medium reinforcing carbon black will give.

These fillers are often incorporated slowly into the rubbers but do not cause heat generation during mixing, unlike carbon blacks. They improve processibility and reduce die swell etc. Except for silicas they may not improve tensile strength. Silicas, silicates, kaolin, precipitated calcium carbonate may increase modulus. Hardness may be increased by those with anisotropy. Whiting and ZnO improve resilience while silica may reduce it. Precipitated silica gives better tear strength than even carbon black.

Non black fillers:

Silica: Silica is the second most important reinforcing filler used in rubber industry. It is of three types – ground silica, fumed silica and precipitated silica.

Ground silica is mildly reinforcing and improves heat resistance for silicone rubbers. Kieselguhr can also be used – this is also a silica from nature – it contains a lot of moisture. It is a semi reinforcing filler for silicone rubbers. It gives acid resistance, lower die swell. They are coarse fillers.

Fumed silica is expensive but pure and of low particle size. It is used for reinforcing silicone rubbers. Unlike ground silica it does not cause silicosis when inhaled (due to its amorphous structure), if densified. The acidic nature of its surface is reduced by treating with triethanol amine or diethylene glycol. Precipitated silica is a hydrated form and hence will contain moisture on its structure. Mixing this into NR should be done with care otherwise they may make the rubber 'dead'. Incorporation should be done fast and without the presence of other ingredients (including plasticizer). They improve mechanical properties of many rubbers.

Calcium carbonate: It is known as whiting. It is basically non reinforcing. They can be mixed easily even at very high loadings like 300 phr. They do not affect cure. Another form is ground limestone – also commonly used in the rubber industry. Precipitated calcium carbonate has lower particle size and hence can be reinforcement. It may speed up cure. A surface treated (with chemically bound carboxylated polybutadiene), precipitated calcium carbonate is also available, which can be fairly reinforcing and hence used for good quality products.

Clays: They are hydrated aluminium silicates. Some are soft clays and others hard clays – soft clays are inert while hard ones are cheap, semi reinforcing fillers. Hard and soft depend on particle size. Clay can be loaded upto 200-300 phr levels. It imparts acid resistance and can be used in products in contact with acids. Recently, silane treatment of clay has been reported to reinforce the rubbers.

Other forms of clay are kaolin – hydrated sodium aluminium silicate got by precipitation – this can be surface treated to reduce particle size and this reinforces many rubbers. Some clays may contain heavy metal ions beyond the desired limits – they may speed up degradation of the rubber product – this must be checked before use in such products.

Hard clay improves modulus and other properties while soft clay does not reinforce much and hence may be used as an inert filler. Clay may be difficult to mix in high loadings – they may come off the rolls during mixing at high loadings. Further they may give poor surface finish for the product if the mix is not softened.

Heating of clay (this clay is called calcined clay) removes the moisture can make it a semi reinforcing filler. This clay has a white colour and is useful in electrical applications as well. Clay is also called kaolin or china clay.

Hydrated aluminium silicate: This is called regenerated kaolin or micro crystalline zeolite. This is obtained by surface treatment of clay. This has a much smaller particle size and is reinforcing. It can be used for all rubbers-especially for footwear applications.

Precipitated aluminium silicate: This is produced by precipitation. This is also used for shoe soles and light coloured consumer goods. For NR it improves extrusion and calendering behaviours without reducing strength. It also improves resistance to steam. At very high levels of loading, it affects cure behaviour due to its alkaline nature.

Precipitated calcium silicate: Natural calcium silicate is not used by rubber industry due to the hard crystals which cannot be eliminated by grinding. Sodium silicates on reaction with calcium chloride leads to precipitation of calcium silicate which is a reinforcing filler. It is cheaper than precipitated silica but reinforcing to medium level. MBT may become unreactive in its presence while other accelerators can be used. It is useful in roller coatings, floor coverings, soles, mechanical rubber goods etc.

Magnesium carbonate: It is obtained by precipitation. Its refractive index is close to that of the rubbers and hence useful for transparent products. It may form lumps and hence may be difficult to incorporate. It improves extrusion characteristics and may increase hardness, modulus and compression set.

Zinc oxide: It was used as reinforcing filler till the advent of carbon black. It imparts good mechanical properties, resilience, flex life, tack etc. It is not used as a filler due to its high volumetric cost (its specific gravity is very high). It is also a curative for some rubbers and an activator for sulphur cure.

Titanium dioxide: It can be a filler but useful only as a tinting material for white products. For silicone rubbers it is a mild reinforcing filler and improves heat resistance.

Barium sulphate: The natural form of this filler can be used as an inert filler and for improving processibility. Its refractive index is closer to that of rubbers. The synthetic form (obtained by precipitation) can reinforce rubbers besides imparting transparency. Lithophone contains in addition to barium sulphate, zinc sulphide.

Talc: Talc (French chalk) is hydrated magnesium silicate with aluminium silicate – a white-greyish white powder, greasy to touch. It is useful in heat resistant mixes, in electrically resistant products, and those with low permeability to gases. They do not increase viscosity of mixes while improving extrusion characteristics. It is also used as anti tack agents in rubber industry while storing compound sheets. Similar uses are known for mica (aluminium-potassium silicate).

Hydrated alumina is used for light coloured products and for flame retardant products.

Litharge (PbO) is an activator and can be a filler providing radiation resistance and water resistance.

Antimony oxide (Sb_2O_3) provides flame retardancy.

Barium ferrite at high loading levels give magnetic properties.

Graphite and MoS_2 impart low fricition coefficient. Lead metal gives resistance to radiation. Asbestos is good for flame and heat resistance. Cork provides resilience and compressibility (for flooring tiles, gaskets etc).

Organic fillers:

High styrene resins: SBR containing high styrene contents (50-85%) are hard polymers with light colour. They do not discolour in light. They impart high modulus, hardness, tear strength and flex resistance while maintaining low density. Main uses are in dirt proof, light and light coloured shoe soles, floorings etc.

Phenolic resins: Novolak type resin with the necessary curative may be incorporated into diene rubbers and during cure, will polymerise in presence of the rubber and also form a few cross links with the rubbers. They can reinforce without increasing hysteresis (heat build up). If loaded in high amounts, they can even form ebonite type compositions with low cure times compared with typical ebonite compositions. Phenolic resin is also a curative for diene rubbers besides being a processing aid.

Plasticisers, Softeners and Processing aids: They improve processibility – in mixing, calendaring and extrusion operations by making the compound more plastic (i.e. lowered viscocity).

Plasticisers are also called softeners – they can be of the following types – petroleum products, coal tar products, natural products, ester types etc.

Among petroleum products, paraffin wax is often used in diene rubbers. It is a hydrocarbon solid with crystallinity. We also have micro crystalline wax. They melt at processing temperatures and soften the rubber, making it easy to incorporate the carbon black. They also act as physical anti ozonant (in static products) by migrating to the product surface and forming a protective layer – this fails in dynamic products as the protective film breaks under such conditions. The micro crystalline wax has branches and rings in its structure and hence it is less crystalline. Due to its higher molecular weight it melts at a higher temperature and hence less brittle. Other waxes are mineral wax (yellow solid and less crystalline), ceresine, petrolatum wax (a semi solid).

Mineral oils: These are most commonly used as plasticizers. They are obtained from crude oil distillation. They are oily substances with molecular weights of about 200-250 – they are basically mixtures of various hydrocarbons which are paraffinic, naphthenic and aromatic in nature.

The fractions can be classified based on viscosity constants. The non aromatic ones may contain aliphatic (paraffinic) and cyclo alkane fractions. The fractions with the lowest viscosity constant will be called paraffinic oils (class A or 0). Class B (also called class I) is relatively naphthenic oil. Class C (also called II) is naphthenic while D (or III) is relatively aromatic and E (or IV) is aromatic. F (also called V) is highly aromatic and G (or VI) is extremely aromatic.

Aromaticity may be determined by aniline point – this is the lowest temperature at which the sample is perfectly miscible with aniline in a 1:1 mixture. The lower the aniline point the more aromatic the oil will be. This is not a perfect measure of aromaticity as the aniline point may also be affected by molecular weight. The end point may be difficult to determine if the oil is dark in colour.

Another important softener is bitumen. Various types of bitumens are known – some are available from nature while others are from distillation of crude oils. They may be semi solids and some grades are obtained by air blowing while distillation is done – such a product is called mineral rubber. Mineral rubber can be ground to a powder and used.

Softeners are also obtained from coal tar distillation. They may be mixtures of hydrocarbons as well as those with oxygen containing functional groups. Some are resins which are very useful in the rubber industry. Brown coal tars are obtained from lignite coal.

Some coal tar distillation products are coumarone resins. They are resins containing coumarone and indene.

These substances polymerise in presence of acid and form a copolymer which may be viscous liquid or brittle solid depending on degree of polymerization. They are used as softener.

Rosin comes from pine trees. The main component in this resin is abietic acid. Due to the acidity it may retard cure. The double bonds present in this also causes quicker ageing of the rubber. Hydrogenation or dehydrogenation (aromatization) can reduce the unsaturation levels and hence solve this problem. Pine tar is another natural resin used in the rubber industry. This contains hydrocarbons, fatty acids, phenols and resinuous acids.

Among synthetic plasticizers, dibutyl phthalate (DBP), di (ethyl) hexyl sebacate, tri cresyl phosphate are commonly used for polar rubbers like NBR.

Esters like ethylene glycol dimethacrylate, trimethylolpropane trimethacrylate are also used – they also get cross linked by peroxide cure and become part of the network thus exerting reinforcement. Halogenated paraffins/hydrocarbons improve flame resistance for some rubbers. Liquid rubbers like depolymerised NR, liquid NBR, poly sulphide etc, are also used as plasticizers – they plasticize during processing but during cure will become part of the network – thus they improve processibility without reducing strength of the rubber.

Phenolic resin is another plasticizer but also a curative for diene rubbers– its extensive use in curing of tyre curing bladders based on IIR is well known. Stearic acid is another plasticizer but also fulfils an additional role of activator of the accelerator.

Mineral oils are the cheapest while the natural ones are more expensive – the most expensive plasticizers are the liquid polymers.

Plasticisers reduce the mechanical properties of the rubber. This is because its presence reduces the rubber content per unit volume, reduces the concentration of the polymer chain entanglements (and cross link density). Plasticiser also provides lubricating action of the oil which reduces internal friction during flow thus reducing viscosity. Plasticiser also reduces elastic recovery after deformation. During mixing, adding plasticizer at suitable stages helps in easy mixing (lower energy consumption during mixing) of fillers into the rubber. Easier wetting of the filler by the oil also helps in compounds which are highly filled by reinforcing fillers. Extrudates show lesser shrinkage, and lesser roughness on the surface. Heat generation during mixing, calendaring and extrusion also decreases due to the plasticizer. Hysteresis and lower service temperature limit, are also reduced.

Though mechanical strength is expected to decrease by plasticizer, it may sometimes improve some of the properties related to strength due to positive effects on dispersion and wetting of the filler into the rubber.

The efficacy of a plasticizer depends on molecular structure, molecular weight and chemical reactivity. Polarity of the plasticizer increases as per the following order – paraffinic < naphthenic < aromatic. The solubility parameter of the plasticizer and that of the rubber should be as close to each other as possible for compatibility. Polarity of the rubbers increases as per the order, IIR < EPDM < NR < SBR < CR < NBR. More compatibility results in faster mixing of the rubber with the plasticizer, wetting and dispersion of additives, and tack. Mixing of resins improve tack of the compound. The restricted solubility of the plasticizer into the rubber causes clusters of lubricated polymer chains move to the layer adjacent to the extruder screw or calendar roll, leading to easier processibility.

Molecular weight affects viscosity and compatibility with the rubber. Lower molecular weight leads to better flexibility of the polymer by increasing segmental mobility. Thus, heat build up, brittle point etc, will decrease. Too low a viscosity/molecular weight will lead to loss of plasticizer by evaporation etc, and hence unavailable after a period of time. More molecular weight of the plasticizer leads to lesser flexibilisation of the compound but also 'kills the nerve', leading to advantages in processing operations like extrusion. The deterioration in mechanical properties due to the plasticizer will be lesser. Solid plasticizers often behave as organic fillers. The overall picture is that, plasticizers of medium molecular weight should be chosen.

Chemical reactivity is also important – any acidic group in the plasticizer will retard sulphur cure of diene rubbers. Some oils contain bases which may accelerate the cure. Double bonds present in the plasticizer may become cross linked with the rubber and hence sulphur should be added in larger amounts. Pine tar is a good plasticizer for NR as it retards cure and also improves wetting of the filler etc, though it may be expensive.

Plasticisers when added upto 16 phr can be considered as plasticizers but above it they should be considered as extenders. For NR, aromatic oils are preferred and used in mechanical goods. Paraffinic oil in NR is mixed slowly and does not improve tack much – they may be used for food grade NR products. BR can withstand large amounts of oils. BR and SBR can be extended with oil effectively. Wax may be added even if plasticizer is added. This improves extrusion characteristics. It also blooms to the surface thus behaving as physical anti ozonant.

Increasing tack is important for SBR more than for NR. Rosin may be added to 1 phr for NR but upto 3 phr for SBR. Phenolic resin also improves tack. Asphalt improves processibility but the compound is stiff on cooling. It may also cross link along with the rubber and hence may increase hardness after cure. Coumerone resin in SBR works like an organic filler. Plastogen, Rheogen and Bondogen are sulphonated paraffinic hydrocarbons of higher molecular weight. They combine plasticizing and peptizing effects. They are added to NR or SBR upto 3 phr level.

For IIR, paraffinic oil is useful. Wax can dissolve in IIR upto 25 phr level. Low molecular weight poly ethylene can reduce stickiness to the mill roll. For EPDM, paraffinic and naphthenic oils are preferred though the latter, to a greater extent. Oils containing unsaturation can interfere with curing by peroxide.

For CR, paraffinic oil is a lubricant, while naphthenic oil is preferrable. It improves extrusion characteristics. Ester type plasticizers improve low temperature properties of CR but cannot prevent crystallization. For CSM, stearic acid and low molecular weight paraffin wax may be useful. For NBR plasticizer choice depends on acrylonitrile content. A combination of esters may work better. Extraction of the softener may be prevented by combining with resins. In ACMs, often the plasticizers may evaporate during post cure. Thus a non volatile softener like coumerone resin must be used. For vitons, some polyesters or low molecular weight poly ethylene may be used. For SBS block copolymers, naphthenic oils are preferred while aromatic oils may swell the styrene blocks and deteriorate mechanical strength.

Factice: They are vegetable oils cross linked by sulphur. They were used as extenders when they were much cheaper than NR but now, since the price of the rubbers have come down they are used only for improving some special properties. Brown factice is got by heating rape seed oil or fish oil with sulphur at about 140-160°C. It can be added upto 20 phi - this improves extrusion characteristics like die swell besides making it easy to mix additives into the rubber. It is also useful in vulcanization in absence of pressure for thin walled products. It also makes the product soft and flexible. White factice is obtained by reacting oil with S_2Cl_2. This is acidic and hence retards cure. It is used in erasers. A high dosage of accelerator is essential to achieve sufficient levels of cure. Its acidic nature may be exploited by mixing in compounds with tendency to scorch. Special factices for use with NBR and CR are also available.

Blowing agents: They are required to make expanded products. The blowing agent usually decomposes and produce gases which cause expansion. The gas evolution must occur when the rubber compound is still in a plastic state. The first blowing agents are inorganic salts like ammonium carbonate or sodium bicarbonate. Sodium bicarbonate forms finer pores while ammonium carbonate forms larger pores. They are dispersed into the rubber with difficulty.

Organic blowing agents are more preferable. They are of the following types: those based on dinitroso penta methylene tetramine (DNPT), those based on sulpho hydrazides (BSH) and those based on azo dicarbamide (AC). BSH is expensive and this leads to DNPT being more preferred. The main disadvantage of DNPT is the fishy odour of its decomposition product which can be suppressed by adding urea.

Its decomposition temperature is about 130°C. BSH and AC have higher decomposition temperature and may be used if higher temperatures are encountered in processing.

Peptizer: It is a chemical plasticizer. Bis (benzamidophenyl) disulphide is a peptizer for NR, SBR and NBR. N,S dibenzoyl 2 amino thiophenol is also used in NR. The most common one is pentachloro thio phenol. They act by combining with free radicals formed by action of shearing on the rubber and thereby stabilizing the free radical. Oxygen is also a peptiser for NR. Some accelerators like MBTS can also be peptisers. Another peptizer is Struktol A.

Anti degradants: These chemicals protect the rubber and improve its service life. Diene rubbers are much more prone to degradation by heat – this is aggravated by oxygen in the atmosphere. The mechanisms of oxidative degradation of rubbers have been studied extensively and this has enabled devising protective systems.

Oxidation of a polymer may pass through the following stages:

Initiation: $RH \longrightarrow R. + H.$ (R is the rubber)   — 1

Propagation: $R. + O_2 \longrightarrow RO_2.$ — 2a

                      $RO_2. + RH \longrightarrow ROOH + R.$ 2b

                      $ROOH \longrightarrow RO. + RO_2.$ etc   2c

Termination: $2R. \longrightarrow R-R$ 3a

                      $R. + RO_2. \longrightarrow ROOR$ 3b

                      $2RO_2. \longrightarrow$ non radical products   3c

These processes are chain reactions. With each propagation cycle, a molecule of hydroperoxide is formed which is the free radical source. Termination may occur by cross linking (as in 3a – this is seen in many rubbers except NR and IIR) or degradation (3c) – as in NR/IIR. These reactions are auto catalytic.

The auto oxidation leading to degradation can be prevented by either interruption of the chain processes or by destroying the radicals as soon as they are formed.

Prevention of auto oxidation is done by chain breaking anti oxidants – they may act by the following mechanisms:

Initiation: $AH + O_2 \longrightarrow A. + HOO.$ (A is the antioxidant)   — 4

Transfer (propagation): $RO_2. + AH \longrightarrow ROOH + A.$ 5a

                      $\overset{O_2}{A. + RH} \longrightarrow AOOH + RO_2.$ 5b

Termination: $RO_2 + A \longrightarrow ROOA$ 6a

                      $A. + A. \longrightarrow A-A$ 6b

In reaction No.4, AH is actually a pro-oxidant. If the probability of 5a is more likely to occur than 2b and the reaction rate of 5b is low, AH can function as an oxidation retarder. Anti oxidants may act by reactions 6a or 6b – these can be called chain breaking anti oxidants.

Peroxide decomposing agents can act as preventive anti oxidants, if the reaction shown below occurs faster than 2c.

$$ROOH + X \longrightarrow ROH + X=O$$

Commercial formulations often do not contain such chemicals – only chain breaking anti oxidants are used often.

Metal ions (especially heavy ones like transition metal ions) can speed up the decomposition of the rubber hydrocarbons. Thus, in NR presence of copper or iron etc, due to the metabolism in the trees, can speed up degradation. They may be prevented by adding chelating agents for such ions. This is not a commercially viable process. UV light can also speed up the decomposition of the rubber hydrocarbon. Luckily, carbon black which is almost an essential additive for rubbers can absorb UV light and to some extent shield rubber products from it.

Due to the presence of double bonds at short distances from each other, the peroxide radical from the rubber may rearrange in a number of ways.

In NR, initial oxygen uptake is rapid and then slows down. The poly sulphide links may absorb oxygen and gets converted to thionates etc, which have been proved to have anti oxidant characteristics.

The type of cross link has a bearing on anti oxidant behaviour of the rubber vulcanisate. TMTD gives mainly mono sulphide cross links and a rubber cured by TMTD shows resistance to oxidation but the byproduct which is zinc salt of dialkyl dithio carbamate (another ultra accelerator) if removed, makes the rubber vulnerable to oxidation – thus it may be concluded that dithio carbamate can promote anti oxidant activity. Further it is well known that poly sulphidic links make the rubber more vulnerable to oxidation – they decompose to mono sulphide links and this leads to increase in compression set values. Peroxide cured compounds have only C–C cross links which are also resistant to set – hence they are preferred for low compression set property.

Anti oxidants may be lost by volatility or by migration or by extraction by the contacting fluid (even water). The good anti oxidants can stain the rubber product. They are the amine types – derivatives of α-naphthyl amine and its derivatives – their oxidation products are coloured and hence stain the rubber – they are also considered toxic. Often the derivatives of β-naphthyl amine (PBNA etc) are used. The unsubstituted one, is a grey powder which furns brown-red. It improves ageing and flex cracking resistance. Its efficacy is better if combined with p-phenylene diamines and its derivatives and waxes. P-phenylene diamines and its substituted forms are good anti ozonants. β-naphthyl amine substituted para phenylene diamine is another good anti oxidant and provides protection against ageing but not against flex cracking.

Similar is the case with poly 1, 2 dihydro 2, 2, 4 trimethyl quinoline. Most of these are added upto 1.5–2 phr levels.

Condensation products of naphthyl amines with aldehydes stain to a lesser extent and can be added up to 3 phr levels.

Substituted phenols are non staining but their anti oxidant ability is not high. Unlike amine types, they do not bloom either. They can be added upto 2 phr levels.

2-Mercapto benzimidazole is another efficient anti oxidant for NR and other rubbers. It retards acidic curatives and slightly speeds up basic curatives. It protects NR against over cure. Upto 0.5 phr it does not cause opacity and hence can be used for transparent products.

In NBR or EPDM some times the anti oxidants should not be leached or lost by volatility as they may face higher service temperatures. For such cases, network bound anti oxidants are considered – they are reacted and grafted into the rubber before they are added in the compound.

Ozone attacks a rubber mainly in the unsaturation sites and when the rubbers is in a strained state. P-phenylene diamine derivatives are often used to protect against ozone.

They can protect even in dynamic conditions. In static products, wax is sufficient as anti ozonant as it blooms and forms a protective layer on the rubber and prevents the ozone from attacking the rubber. In dynamic conditions this layer breaks and exposes the rubber to ozone. More saturated rubbers/ polymers are naturally resistant to ozone – like butyl or EPDM and hence blended to SBR or NR in some applications. They may be considered as polymeric anti ozonants.

Reclaimed and recycled rubber:

Reclaimed rubber is another cheapening additive for many compounds. They were initially required for countering the shortage of the vital raw material NR (which was controlled by Britain) which was much needed in the early part of the 20th century for making tyres. Earlier natural and synthetic rubbers (SBR and BR) needed separate reclamation processes. Now a days, technology common for both, is required, as blends of these rubbers are more common.

Prior to reclaiming, ground scrap tyres too were used as additives in various rubber compounds. Their uses are today, confined only to asphalt modification for road laying-this improves the life of the asphalt top on the roads. Reclaiming of a rubber is a better option. Grinding may be done in cryogenic temperatures when the rubber becomes brittle and hence can be ground to a powder. This is a costly option as cryogenic temperatures are costly conditions to attain. Cryo ground rubber can be added to an extent of 10% by weight of virgin rubber in its formulations. Further additions make the compounds too stiff.

Reclaiming is important for two reasons – to mitigate the problem of pollution caused by pile up of used tyres and for producing cheap raw materials at least for non critical applications.

The first step in reclaiming is to remove the non rubber constituents of the tyre – bead wires and fibres. Bead wires are removed before the tyre enters the reclaiming area. The tyres are cut to smaller pieces and sent to equipment like cracker mills (i.e. mixing mills with grooves, to facilitate the cracking) after passing through mills with grooved discs. The fibres are removed by sieving or pneumatically. Still a small amount of fibre may remain – they may removed by digestor process.

Reclaiming may be by steam or by digestion or thermo-mechanical.

Steam process involves heaters of pans. The ground tyre is mixed with reclaiming oil in a mill and put into open pans and heated in live steam in the pans. The treatment requires about 3-5 hrs. This method was replaced by steam at higher pressure (static steam reclaiming). Further, this too was replaced by dynamic steam reclaiming – reclaiming is done while the material is moving.

Digestor processes involve the coarser rubber particles being suspended in water, in the presence of acid or alkali and heating under live steam. In these methods, fibre separation is done simultaneously by the acid or alkali and also heat transfer is uniform. The digestion is done in autoclaves. NR tyre crumbs require alkaline digestion while tyre crumbs containing synthetic ones, acid or neutral digestion.

Thermo mechanical methods involve working up the scrap in banbury type mixers – here temperature in the chamber increases to 220-290°C. Softening is done in 3-5 mins – the heating is uniform compared with other methods. Later equipment are of screw – extruder types. The shear heating as the material passes through the barrel causes the devulcanisation.

The material which comes out may be of powder form which is often made into slab forms after homogenization in two roll mills.

Reclaimed rubber is an important additive – it reduces compound cost, saves energy for mixing compounds, improve extrusion and calendaring behaviour (reduced die swell, reduced shrinkage and better shape retention of the extrudate etc), reduce heat generation during processing operations, Reclaim speeds up cure while decreasing the heat generation during cure and hence added in ebonitic compositions. Ageing resistance is also increased by reclaim addition. Tackiness is also improved. The rubber based adhesive industry uses reclaims commonly in its formulations.

In tyres, reclaim can be used in carcass compounds in bias tyres, and in bead compounds but not in tread formulations as it deteriorates the properties drastically in this case. Conveyor belts, shoe soles, automobile floor mats, battery boxes, etc, are other products using reclaimed rubber.

It must be remembered that reclaiming process causes mainly depolymerisation rather than devulcanisation.

Microwave devulcanisation: Microwaves can be used to cleave C–C bonds in the vulcanisate – this mainly devulcanises the rubber rather than depolymerise it. When added to virgin rubber the mechanical properties attained are as good as those of original rubber. For this technique to be usable the rubber compound must contain polar links – the sulphur cross links fulfill this need. Similarly, ultrasonic waves too can be used to devulcanise rubbers. However, they cause considerable depolymerisation and hence properties attained will be poorer.

Devulcanisation by chemicals: Some dialkyl aryl amine sulphides can break C–S–S–C cross links and lead to devulcanisation. Ground scrap may be placed in trays and treated with such chemicals and heated in autoclaves – steam pressure of about 4 bar and good circulation of air and steam will be needed. Later, higher pressure steam should be applied till the temperature reaches 190°C. After about 3-5 hrs, the material undergoes devulcanisation considerably.

The mechanisms may be similar to the reactions leading to change in sulphur ranks in poly sulphide rubbers. Initially such reactions were used for breaking cross links for studying the curing mechanisms of rubbers – to assess the extent of formation of mono, di and poly sulphide links in the network.

Thiol-amine systems, triphenyl phosphine and lithium aluminium hydride can break poly sulphide links while tri phenyl phosphine can also break disulphide links. Mono sulphide links can be broken by methyl iodide.

$$R' \, SH{-}\overset{\displaystyle \backslash}{\underset{\displaystyle /}{N}}H + R{-}S{-}S{-}SR \longrightarrow R'{-} \, S{-}SR' + RSH + HSN^{+}HR'$$

$$Ph_3P{:}R{-}S{-}S{-}R \underset{\longleftarrow}{\longrightarrow} Ph_3P^{+}SR\} \, SR^{-} \xrightarrow{\,H_2O\,} Ph_3P{=}O + H^{+} + RSH$$

Dr B.C. Sekhar from Malaysia has discovered a chemical called Devulc which can be mixed with rubber scrap and give devulcanised rubber – mainly for coloured products.

Some methods have also been derived to obtain monomers or oligomers by pyrolysis of used tyres – till now only a very few can be commercialized.

7

Thermoplastic Elastomers

The rubber industry has been in search of materials which can be processed and recycled like a thermoplastic but still maintain rubber-like elastic properties. This is because of the three problems of scorch, inability to use production scrap and disposal of rubber waste. These are the main disadvantages of rubber processing and usage. Further, the rapid automated production systems characteristic of thermoplastics – i.e. melt processing techniques, continuous mixing techniques, forming, blow moulding etc, are not available for the conventional rubbers.

Thermoplastic elastomers (TPEs) can solve these problems. Overall, the advantages of TPEs are as follows:

1. Faster processing

2. No scorch danger

3. No need for reinforcing fillers like Carbon black and its accompanying problems

4. Lighter machinery and moulds compared with conventional rubber for the same level of mass production

5. Scrap and failed products can be recycled by simple remelting etc

6. Applicability of melt processing techniques like plastic injection moulding, blow moulding

7. Applicability of thermoforming

8. Ability to produce aesthetically appealing products which is not easy in conventional rubbers.

The disadvantages of TPEs are:

1. Poorer heat and swelling resistance

2. Poorer set properties

3. Elastic properties are inferior to those of conventional rubbers

Further, newer design considerations have come to play with the advent of these materials – for example, an assembly for use in automobiles which was previously made of a combination of rubber and metal components can now be moulded as a single piece using a TPE.

Another interesting thing is, in the past TPEs were considered as a threat for rubbers only but now a days, TPEs are also replacing plastic components due to their unique set of properties.

Various approaches towards TPEs:

TPEs can be developed by any technique which can lead to a polymer system where hard segments (blocks) are connected to each other through soft (flexible) segments. This system may be likened to a situation where hard spheres are connected to each other by means of springs.

In a typical elastomer, the material can undergo flow under pressure and heat. This is arrested by vulcanization. To improve strength, reinforcing fillers are added. They adsorb the rubber on its surface and some rubber gets 'bound' on to the surface – especially in carbon black (in silica filled compounds this role is facilitated by coupling agent). They also restrict the deformability of the rubber and this contributes to strength.

The hard blocks in a TPE has to fulfill the roles of both the chemical cross links and also the reinforcing fillers. However, the cross links are physical in nature and can disappear on heating and reappear on cooling.

For the kind of molecular architecture leading to TPE, the following approaches can be of help:

1. Block copolymers where one block is a hard (plastic) block and the other one, is soft (elastomeric) – this can be observed in styrene-butadiene-styrene (SBS) or styrene-isoprene-styrene (SIS) or styrene-ethylene butylene-styrene (S-EB-S) block copolymers or poly urethanes or polyester or poly carbonate based block copolymers.

2. Ionomers – e.g. polyethylene ionomer

3. Blend of a plastic with plasticizer e.g. Plasticised PVC

4. Blends of a plastic with a rubber (NR + EPDM or PP + EPDM)

The block copolymer route was the first approach. In these, one block must be rigid and the other, flexible. The rigidity can come from a high Tg polymer (as in SBS etc, where the Tg of styrene block is +100°C) or from crystallinity as in polyester TPE or from H bonding – as in PU or poly amide TPEs. Similarly in plasticized PVC, H boding between the polymer and the plasticizer contributes to strength. It must be remembered that the PVC-plasticiser attractive forces are high and this leads to a situation where a small amount of plasticizer in PVC, instead of plasticizing actually causes an anti plasticization effect – i.e. addition of plasticizer increases viscosity – this can also lead to a rigid block formation leading to TPE behaviour.

In ionomers, like polyethylene which is copolymerized with acrylic acid, the COOH group in the ends can be cross linked by metal oxides and form ionic cross links. These cross links make the melt elastic. Entropy-wise, the ions cannot be found dissolved in a non polar polymer mass and hence must try to associate with each other and hence form clusters, as the melt is cooled. This becomes the hard block. On heating, the clusters will temporarily break and the order gets disrupted – hence the cross links are heat – fugitive and thus the material is thermoplastic.

$$\ce{\overset{\diagdown}{COO^-} \ Zn^{+2}}$$
$$\ce{\underset{\diagup}{COO^-}}$$

$$\ce{\overset{\diagdown}{COO^-} \ Zn^{+2}}$$
$$\ce{\underset{\diagup}{COO^-}}$$

The poly ethylene chains between the ionic clusters will act as the elastomeric phase. Poly ethylene ionomers are tough materials with good properties – some of their uses are in golf balls. Other uses are packaging films. Ionomers also play a major role in developing compatibilisers for blending dissimilar polymers.

Ionomers based on other polymers like PU are also available.

SBS type block copolymers: Block copolymers of this type can be manufactured by anionic polymerization exploiting the concept of 'living' polymers. On initiation of a monomer to polymerization by anions, the growing chains do not have a pathway by which the reactions can be terminated, unless an external reagent is added. In case of free radical polymerisation, two radicals can destroy each other by combination or bimolecular disproportionation. Thus one can polymerise styrene to the desired molecular weight, then without destroying the anions, the other monomer can be added (here it is butadiene)- now the styrene blocks will not have the ions but the butadiene blocks which are connected to the styrene blocks will have the negative charge at the other end of the chains. In this condition again styrene solution may be added – once again the negative charges will transfer to styrene which will now polymerise and have the charges on the other end. Now the reaction may be terminated. Thus we can get S-B-S polymer in this sequence – this polymerization technique is called sequential addition polymerization. By changing the order of addition we can change the type of polymer like SB or BSB polymers. Now a days, other methods are available for preparing block copolymers – like the use of bifunctional initiators like dilithio dialkanes, which can lead to formation of star shaped polymers etc. Free radical methods are also possible but generally have poor control over the polymer structure and hence not preferred. Co-mastication of two polymers also can give block copolymers by free radical route.

Properties of block copolymers:

Considering the three types of polymers obtained by sequential addition polymerization, we compare their properties. Of these, only SBS works as a TPE while SB and BSB do not – they can be considered only as toughened plastics. This is because, in the SBS block copolymers, the middle (elastomeric) block consists of rubbery chains which are embedded into end (plastic) blocks – thus we can see that this system behaves like a rubber – the hard blocks fulfilling the roles of chemical cross links as well as those of reinforcing fillers. The cross links in a conventional rubber prevent flow while the reinforcing filler reinforces the rubber by adsorbing the rubber chains and immobilizing a few segments and arresting the chains from elastic deformation under an applied tensile load but dissipating the strain energy and thereby preventing any possible flaw from spreading (this leads to failure).

Such a system is not possible if the rubber is the middle block or if it is not bound by plastic blocks on both sides. Thus only SBS is elastomeric while the other two are not.

These materials if viewed under electron microscope, will show two phase behaviour. The continuous phase will be the one whose content in the copolymer is high while the other component being the discontinuous phase. It is possible to vary the relative compositions and the molecular weights of the two blocks. Comparing two block copolymers where one is SBS with the molecular weight of styrene blocks being 10000 each and that of butadiene block being 52000 while another being BSB with molecular weights of the styrene and butadiene blocks being 21000 and 28000 each, respectively, the first one behaves like a rubber with good tensile strength, elongation at break and hardness (65 Shore A) while the other one shows very poor tensile strength, elongation at beak and hardness of about 17 Shore A units (It must be noted that the styrene content and the overall molecular weight was the same in both the cases). For a good TPE behaviour, a typical SBS polymer has about 70% butadiene with the molecular weight of this block being in the range of 50000-1,00,000 and about 30% of styrene with each styrene block having molecular weight around 10000.

TPE-may be pictured as spherical domains of PS (rigid balls) connected to each other by Polybutadiene (elastic spring) segments.

Morphology and transitions: In the typical TPE compositions, the butadiene contents are higher – this means the butadiene block will be the continuous phase while the styrene phase is the discontinuous/dispersed phase. For a polymer with molecular weight of poly styrene block being 21000 and poly butadiene block being about 64000 and with styrene content being about 38%with one can see styrene blocks dispersed as spheres into a poly butadiene matrix. Some kind of ordered arrangement (but nowhere near crystallinity) is seen, with the styrene blocks having a domain size of about 350 A° units and with the inter domain distance of about 560-670 A°.

If the elastomer content is very small (i.e. below 20%), it will be dispersed as spherical domains (of a few hundred A° in a plastic matrix. If the elastomer content is increased, the spheres will not grow in size beyond a point but become cylinders – this occurs when the styrene content is between 20 and 40% approx). Between 40 and 60% styrene content, lamellar arrangement will be seen – alternate "sheets" of poly styrene and poly butadiene blocks. For a typical TPE, the spherical domains of the plastic embedded into a continuous, rubber phase is the optimal arrangement.

This morphology can change by contact with solvent or by shearing. A specimen may come in contact with a solvent – the solvent may be able to dissolve both the blocks or only one of the blocks. For SBS, a solvent capable of dissolving both the blocks may cause phase mixing and this may give rise to a Tg of the polymer being intermediate between those of the pure PS and PB blocks. On shearing – subjecting the material to flow, the spherical domains may become oval in shape eventually. In the initial stages, the PS blocks align perpendicular to the direction of the applied force and they further along with the stress field forming V shaped patterns with the spheres getting connected to each other. Corresponding to this, stage, the stress required to have this deformation will be high. On further elongations, the connections between the spheres will break and the material will now undergo creep – this also dissipates large amount of strain energy. After the spherical domains become oval and the stretching is continued, ultimately, failure will occur.

The extent of phase separation also has a bearing on the mechanical strength of the TPE. If the phase separation is more marked, the strength will be better while if the phases are more compatible with each other, strength will be poorer. Thus, SBS is stronger than SIS copolymer.

Mechanical properties: These materials are very much like vulcanized rubbers – tensile strength can go up to 30 MPa while elongation at break up to 800%. For a constant PS content, the tensile strength and modulus do not vary much with molecular weight as long as the molecular wt of the PS block is high enough. They also show Mullins effect like a vulcanized rubber (i.e. stress softening).

Viscosities are higher than those of the similar random copolymers or the homopolymers – this may be due to more energy required to pull the butadiene blocks into the flow. If the compatibility between the phases is poor, the increase in viscosity will be more pronounced – as with S-EB-S compared with SBS or SIS.

Substituting poly styrene block with poly (α-methyl styrene) block gives better strength and higher upper service temperature limit as its Tg is about 150°C. However, the middle blocks are not able to withstand the higher processing temperatures and hence this situation cannot be exploited. One solution to this problem may be to use poly dimethyl siloxane instead of poly butadiene or poly isoprene as the middle block.

One practical application of the visco elastic properties of these materials is to use tackifying resins compatible to both blocks – some of them dilute the elastomeric block while modifying the visco elastic response – this is useful in hot melt pressure sensitive adhesives.

TPEs based on PU (Poly Urethanes): The first PU based TPE was discovered in BF Goodrich Co., USA in the 50s. Since then a number of producers have been manufacturing this material in various grades and different properties. The applications of these materials are numerous.

The hard segments (blocks) are formed by the reaction of chain extender and diisocyanate – mainly diphenyl methane diisocyanate (MDI). The chain extenders are often butanediol and hydroquinone bis (2-hydroxyethyl) ether. These can react with diisocyanate to give crystallisable blocks. Diamines as chain extender give hard segments which melt at temperatures well above those for TPE processing.

The polyols and diisocyanate react together to give soft blocks. The polyol may be polyester polyol or polyether polyol. The polyester polyols may be made by condensation raction between adipic acid and diols which give crystallisable soft segments which melt at about 50-60°C. Other polyester polyols of commercial interest are poly caprolactone.

Poly carbonate type polyols are also available which give better hydrolytic resistance.

Poly ether polyols give better chemical resistance and hence TPEs based on them are preferred in some areas. Polyether polyols may poly (oxy propylene) glycols or poly (oxy tetra methylene) glycols (also called poly tetra hydro furan). Hydroxy terminated poly butadiene is also used some times – mainly in propellant binder applications. Soft segments may crystallize only on stretching to medium levels.

Stoichiometry: Ratio of isocyanate to hydroxyl (+amine) groups – combined total of polyol and chain extender whould be in the range 0.96 to 1.1.

Below 0.96, low molecular weights will result while above 1.1, processibility becomes difficult. Often the prepolymer route is used and the reaction is done in reaction extruders. Heat history during the manufacture has a bearing on the properties.

Morphology: Here too, hard domains are connected to each other by soft segments as in SBS types. However, the hard segments here, are more ordered (detected by X ray diffraction studies) than those in SBS and they can be called para crystalline materials as some ordering is observed but not ordered enough to be called crystalline.

Two types of crystallization may be observed – two different transition temperatures may be seen. Another important difference compared with SBS is, the dimentions of the domains are smaller compared with SBS – good properties are obtained with lower molecular weights in PU compared with SBS TPEs. This is due to the crystallinity and H-bonding in the hard domains. Annealing affects properties – as slow cooling leads to formation of better crystals in the hard segments – hence better properties on annealing. Degree of phase separation is determined the Tg of the soft segment. Hardness may be increased by increasing the hard segment content. This in turn increases the extent of phase mixing – this decreases the low temperature flexibility. This can be countered by incorporating soft segments of higher molecular weights.

Good mechanical properties can be obtained by using hard and soft segments of lower molecular weight ranges compared with SBS. Narrow segment size distribution and perfection, in the hard segments leads to better mechanical strength. Broader distribution leads to better set properties.

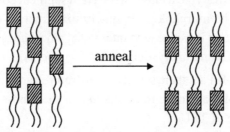

Hardness varies from 70 Shore A to 80 Shore D values. Tensile strength from 25 to 70 MPa. Their abrasion resistance is outstanding. Tear propagation resistance is also excellent. Such properties are shown by TPUs based on poly ester polyols while poly ether polyols give better low temperature flexibility and better chemical and microbial resistances. TPUs based on poly ester polyols will be attacked by acids and alkalies on prolonged exposure. Resistance to non polar solvents is good.

These materials are amenable to processing by the usual thermoplastic machinery.

They can be blended with other polymers. With epoxy resins or poly acetals or PBT, they may act as tougheners. For PVC it is a permanent plasticizer.

TPU may be added to equal parts of poly carbonate – this improves processibility without reducing modulus. When added to ABS at equal parts, it gives better higher modulus.

With TPU as the major component in the blend, ABS improves impact properties, acrylics improve processibility.

Uses: TPUs are used in hoses, shoes, ski boots, ice hockey boots, exterior automotive parts, bellows, shock absorbers, seals for door locks, couplings, precision cogged wheels, sheathing for cables. The bio compatibility is good – hence they are used in catheters, tubes for blood, vascular prosthesis etc.

Polyester type TPEs: They are also called polyether-polyester type TPEs. They are prepared by reacting poly tetramethylene ether glycol (PTMEG), dimethyl terephthalate and butane diol. PTMEG and dimethyl terephthalate react together to give soft segments while dimethyl terephthalate and butaene diol react to give hard segments.

The molecular weight of PTMEG is usually in the range 600-3000. Prepolymer of dimethyl terephthalate with PTMEG and butane diol is prepared. The ester interchange is completed, biproduct methanol is removed and then further heating (to about 260°C) is done at low pressure to remove the excess diol added in the reaction mixture initially.

Molecular weight of this elastomer is often 25000-30000 which is very low compared with the typical elastomers. Hard segment content can be varied from 33 to 85%. The morphology is spherulitic structure consisting of hard segments in the form of lamellae with inter radial amorphous regions consisting of the soft (PTMEG) and uncrystallised hard segments.

Compared with PUs their service temperature range is a little wider as the melting point of the hard segments is higher than those of TPU. The uses are similar to those of TPUs.

Hardness values range from 35 to > 75 Shore D units. Melt viscosities are lower than those seen in TPUs and are less sensitive to shear rates. Due to their low melt viscosities they can be used in processes like calendaring, melt casting, rotational moulding and impregnation into porous substrates. Blow moulding requires special grades – as the standard grades will have low melt viscosities, low melt elasticities and low die swell. Grades with higher molecular weights or those with chain branching, etc, are suitable for blow moulding.

Compounds require the presence of UV stabilizers and anti oxidants. They may be prone to hydrolysis by water at high temperatures though to a lesser extent than PUs. Melt stability is needed for processes like melt casting – this is provided for by poly epoxies.

Another type of poly ester TPE is a poly carbonate TPE obtained by phosgenation of a mixture of PTMEG and a highly cyclic glycol. They can be made to strong elastic fibres. This TPE is not known to be commercially exploited.

Polyester TPEs can be blended with elastomers like NBR and EPDM. besides platicised PVC.

Polyamide type of TPEs have also been prepared and some grades have come to the market. They are of the types – poly ether amides and poly ether ester amides. They can be prepared by reacting MDI with the appropriate polyester or poly ether pre polymers. Each step yields two CO_2 molecules.

In these TPEs the hard segments are crystalline. The tensile properties depend on the amide content in the polymer chains. The values of tensile strength can go up to 40 MPa and hardness upto 70 Shore D units. Their high temperature performance can be in between those of TPUs and silicones. They may withstand upto 150°C – this surpasses the upper limit of TPUs.

Many of the applications will be similar to TPUs but with higher temperature performance requirements.

TPE by blends: This is the most promising development in the field of TPEs. Often it is easier to blend two existing polymers and evaluate their performance rather than synthesise a new polymer.

One of the earlier TPE by blend is the NBR-PVC blend. As early as 1938, a blend of PVC with NBR was reported. NBR with ACN content less than 28% were found to be incompatible with PVC. The resulting blends were of poor properties. Those with ACN content of 35–40% were found to be the best for blending with PVC – they gave the best properties as well as better processibility. With ACN content of 40% the blends were almost homogenous.

The blending was done by mixing was done followed by fluxing at high temperatures to cause more intimate mixing between the phases.

NBR–PVC blends are not used much as TPEs except for a few applications. Often, cross linked thermoset compositions are used in the market – LPG tubes, hoses etc.

Similarly, EVA-PVC blends are successful in a few areas – a small amount of grafting of the vinyl chloride onto the EVA ensures better compatibility in these blends. Some uses may be in transparent hoses/tubes etc, for various applications.

Poly olefin blends:

Butyl rubber can be blended with PE or PP. A few blends were successful in the 70s. Now a days the more successful ones are obtained by blending

EPDM with PP. Monsanto company, USA was successful in this field and many grades were introduced by them. Two types of blends are possible – simple physical blends and those containing a few cross links. The strength is provided by the crystallinity in the PP/PE phase while the elastomer acts as the soft block as in SBS type TPE.

The blends often contain finely divided elastomeric particles dispersed in a plastic matrix. The elastomeric particles must be cured – this is done by Dynamic Vulcanisation. This term means that during the melt mixing of the rubber with the plastic, cross linking in the rubber phase occurs. Dynamic vulcanization results in reduced set, improved elongation at break, improved fatigue resistance, better solvent resistance, better high termperature performance, better stability of the phase morphology in the melts during processing, better melt strength and better thermoplastic fabricability. It is obvious that they have better high temperature performance than the simpler block copolymers. A reason may be the rubber particles forming a network during the melt processing. During regrinding and reworking these will break and his restores thermoplastic nature.

The blending can be done in Banbury mixers or extruders – often twin screw extruders are preferred. The elastomeric phase (often in smaller amount in the blends can be extended by plasticizers. In the molten state they plasticize the plastic phase while on cooling they migrate to the rubber phase – hence it may act as plasticizer at higher temperature while as a flexibiliser on cooling. Mixing is continued for a while after the mixing torque passes through a maximum – this improves thermoplastic processibility.

The dynamic curing may be effected by sulphur or peroxide or other systems. Peroxide systems may cause degradation and hence dscolouration, of the PP at the melt temperatures – this often leads to sulphur cure as the choice. Similar is the case with NR-PP blends (thermoplastic NR, developed in Malaysia). Strength is maximum at about 50 phr of PP content beyond which the values do not decrease but the compositions become less elastomeric. The particle size of the vulcanized rubber inside the PP matrix is about 1.5 micron – this means low amount of cross linking is enough. Higher particle size (caused by increase in cross link density in the rubber phase) leads to lesser values of strain – to – failure values.

Other thermoplastic-rubber blends:

NBR-PP blends: They are expected to give excellent solvent resistance due to the elastomer being polar and the plastic being crystalline (PP) material. The impediment was the large solubility parameter difference between the two polymers. This was solved by dynamic curing – by phenolic resins. The resin reacts with unsaturation sites in the same way as the resin cure of butyl rubber.

PP often does contain a very few double bonds in the backbone and they help in this compatibilisation. The procedure is to react PP with the resin and heat in the extruder so that the reaction occurs. Later NBR is mixed and worked up, to compatibilise the blend. Often one end of the resin is grafted to PP while the other, to NBR. A small amount of ATBN improves the performance of these blends. Instead of phenolic resin, maleic anhydride grafting onto PP can also help in compatibilising the blends.

Phenolic resins

These blends provide good properties but lack low temperature flexibility. This may be improved by blending in a small amount of PP-EPDM TPE.

Another interesting blend is the nylon-NBR blend. They are also prepared by melt mixing. Phenolic resin may act as compatibiliser. Reaction between the resin and nylon may increase the viscosity of nylon to a high value and this facilitates melt mixing of nylon with NBR.

Uses of the blend type TPEs may be in the field of automotive areas – like hoses, bellows, seals, profiles, mounts, bumpers, shields, valves, torque couplings, connectors, handles, grips, steering gear boots, emission tubing, air ducts, hydraulic hoses, mine hose, insulations, jacketing etc.

8

Blending and Compounding of Rubbers

By now it must be clear that hardly any rubber can be used without compounding for any product. Compounding imparts better properties to the rubber without which the uses of rubbers would be severely limited.

Compounding has three essential factors – improvement of properties, ease of processibility with minimal cost implications. The second and third factors are equally important without which a compound designed, will fail.

Compounding is partially a science and partially, art. The compounder must draw conclusions from physical properties. The ability to select appropriate physical properties for a given service is his most important skill.

The compounding ingredients may be classified as: elastomers, cure systems, fillers, processing aids, and miscellaneous ingredients.

The most important ingredient is the polymer itself. Normally we select the lowest cost polymer with satisfactory performance requirement for this purpose. Another important criterion is the processibility. Comparing NR and SBR, these two elastomers are competitive in cost and performance – the choice depends on the relative cost – e.g. in Asia, NR may be preferable while in the West, SBR is the choice. NR will be preferred if high tensile strength, high elongation at break, fatigue crack propagation resistance, tear strength, low heat build up and building tack are required while SBR will be preferred if better ageing and reversion resistance are needed.

For oil resistance, polarity is the factor while for flame resistance, presence of halogens in the rubber is the major factor. Set properties depend on cross link density and any rubber which cannot be cured to a good level cannot have good set properties.

For a combination of heat and solvent resistance, NBR is not enough and hence we need ACMs and for still higher temperature service, FKMs, KALREZ etc, are required. For bio compatibility and service at extremes of temperature

(–100°C to 250°C) silicones are the only choice. It must be remembered that in modern automobiles, under the hood temperatures are increasing, and this necessitates the use of oil resistant rubbers with better heat resistance than NBR. Thus ACMs, EMA and HNBR are increasingly replacing NBR in such applications.

For shaping in liquid state, PU, silicones and poly sulphides are suitable. For extreme solvent resistance poly sulphide rubbers are good. For contact with very oxidizing chemicals, fluoro elastomers are the only choice.

Compounding for specific requirements:

Processing properties:

Viscosity control: Raw polymer viscosities are measured using Mooney viscometer. This equipment operates at very low shear rates compared with the actual processing conditions and also at 100°C which is again below the actual processing conditions. Still this equipment is used for comparing the viscosities. For easy mixing, easy blowing (in sponge compounds) and also for high loading of fillers an initial low raw polymer viscosity (around 30 Mooney units) is desirable while for economical compounding (where high oil loadings are desirable) and also for good green strength as in extrusion products, high Mooney values (100 units) are desirable. High viscosity can be achieved if polymers with temporary cross links as in branched grades of solution SBR are used – these break during mixing and provide easy mixing while retaining high initial viscosity. Oil loading can reduce viscosity while deteriorating mechanical properties while highly reinforcing fillers increase properties, improve extrudability but at the same time increases heat generation during mixing – this may cause "over mixing" and deteriorate mechanical properties.

Adding liquid rubbers in unsaturated rubbers – like liquid NR in NR compounds improve processibility while not deteriorating mechanical properties because they become part of the vulcanisate network.

Nerve – elastic recovery during processing but before cure, should be controlled – for this, use of factice or high structure blacks, or higher loading of fillers, will be of help besides choosing polymer grades which have temporary cross links. These also help in improving extrusion or calendaring behaviour of compounds (like dimensional stability during extrusion, reduced die swell).

Tack and adhesion to mill roll can be improved by adding tackifying resins. Avoiding excessive filler loading, waxes, steric acid etc, in the compound, can help in avoiding drying of the stock and hence promote tack and adhesion to mill rolls. Stickiness to the mill roll or other processing equipment can be reduced using stearic acid or wax in the compounds.

Continuous vulcanization: For continuous vulcanization, compounds should be free from volatile additives and those containing moisture, as the cure temperatures will be high. Further rapid cure is preferable as this will avoid air pockets in the mix which will expand during cure and cause blisters. Similar considerations hold good for injection moulding due to the high temperatures involved.

Scorch can be avoided using pre vulcanization inhibitors (in case of sulphur cure) or slow curatives in case of other cure systems.

Compounding for vulcanisate properties:

Hardness and modulus: Carbon black and other particulate fillers increase hardness. For every rise of 1 unit of hardness rise (in IRHD scale), different rubbers need different amount of fillers e.g. if we take ISAF black, CR requires 1.3 phr of the black, IIR, 1.5 phr, NR 1.7 phr and SBR 2 phr. Similar is the trend with other grades of carbon black. The trend is not so clear for non black fillers.

Hardness is a modulus measurement at very low strain while modulus measurements are done at higher strains (minimum 100% and maximum at 300%). Thus, compounding which increases hardness will also increase modulus. Further, coupling agents also increase modulus in case of silica filled compounds but not necessarily hardness. Coupling agents improve polymer-filler interactions leading to improvements in properties without reducing resilience.

Modulus and hardness are improved by fillers and increasing cross link densities – once again the former method is preferred as it is easier to monitor. Hardness shows a similar trend. Often compounds are evaluated by the users of elastomeric products, by hardness alone but this can lead to disastrous consequences. The simple reason for this is that if hardness alone is the criterion, this property can be improved simply by adding a cheap filler. This will be accompanied often by deterioration in other properties.

Phenolic resins or reactive monomers like acrylates improve processibility during mixing but during cure, forms cross links and increase modulus and strength and hardness. Non reactive resins like high styrene resins (in NR,SBR) and PVC (in NBR) too improve hardness.

Among accelerators sulphenamides give slightly harder vulcanisates compared with thiazoles. Boosted accelerated systems give better hardness than straight systems.

Elasticity: Elasticity may be measured as resilience or set properties. Reinforcing fillers generally reduce resilience and worsen set properties. It is well known that more efficient vulcanisation systems give better set properties but poorer mechanical strength. Plasticisers improve resilience when loaded up to certain levels beyond which they may not improve resilience. In SBR, NBR and IIR the grades containing more of the diene will have more resilience.

Reinforcing fillers reduce resilience – however if polymer-filler interactions improve by using chemicals like coupling agents, increase in strength may be accomplished without loss of resilience – as in case of silica – silane coupling agents in tread formulations.

Tensile strength: Very few rubber products fail by tension. Still tensile strength is a very important measurement for mechanical characterization because it is sensitive to even small variations in compounding and also variation in tensile strength can show variations in the other failure properties. Tensile strength may be improved by using reinforcing fillers like HAF black. The tensile strength vs filler loading curves show a maxima with SBR, NBR etc, but not with EPDM (or other low unsaturation rubbers). Strain induced crystallization in NR, CR and IIR make these unfilled rubbers show considerable good tensile strength.

To some extent, increasing cross linking density may improve tensile strength but not so efficiently as filler loading. For good tensile strength values, particle (filler) dispersion must be fine. Reactive resins, coupling agents (for silica filler) and reactive monomers also improve strength.

Abrasion resistance: Abrasion resistance depends on conditions of wear and hence it is difficult to suggest compounding techniques for improving this property. Blending BR into NR or SBR and adding reinforcing fillers (i.e. those with finer particle size) may improve abrasion resistance. Some times, even adding anti oxidant or coupling agent (in silica filled compounds) may improve abrasion resistance.

Tear resistance: For tear resistance, strain induced crystallization can be a criterion – NR, CR are better rubbers for this property. In NR and IIR, carbon black particles improve tear resistance but not in other rubbers. Some times in SBR coarser blacks were shown to improve tear resistance if any chemical which can improve wetting of the filler by the rubber is added – this may be because coarse particles may divert the tear propagation path through a more tortuous path which may consume more energy for propagation.

For fatigue resistance, crack initiation resistance and crack propagation resistance must be considered – for the former, lesser unsaturation elastomer (SBR rather than NR) will serve better (here too anti oxidants improve this property) while for the latter, a polymer undergoing strain induced crystallization will serve better (for this NR is better than SBR).

More poly sulphide cross links in the vulcanisate can improve fatigue life.

For lower heat build up, rubbers with higher resilience will be suitable though BR sometimes defies the rule – this has been explained in an earlier chapter. Adding plasticizers or reducing reinforcing fillers may reduce heat build up.

Heat resistance: Choice of elastomers for heat resistance has been described in the chapter on structure-property relationships. Rubbers with lesser or no unsaturation sites like EPDM, IIR, silicones and those containing fluorine, silicon-oxygen main chains all have excellent heat resistance.

Among curatives, conventional cure (high sulphur and low accelerator dosages) gives mainly poly sulphide cross links gives better mechanical properties and elasticity (resilience and flex resistance) but poor heat and set resistance while efficient cures (low sulphur and high accelerator) gives better heat resistance and set properties but poor mechanical strength and resilience. Peroxide cure also give excellent heat and set resistance.

Anti oxidants also contribute greatly to resistance to heat ageing. Those which are chemically grafted to the rubber can improve the life as this guards against loss of anti oxidants by leaching or evaporation etc.

Among plasticizers, paraffinic oils give better heat resistance while aromatic ones deteriorate it. Blending EPDM into NR or SBR or NBR can improve their heat resistances.

Flame retardancy: Rubbers containing halogens are self extinguishing e.g. CR or FKM. For further improvement in this property, halogenated plasticizers, antimony tri oxides or hydrated alumina as fillers, will be useful. Halogenated plasticizers will form halogens on heating which will settle down and cover the flame there by cutting off oxygen supply to the flame. Hydrated alumina will release water on heating which will cool the burning area and prevent burning.

Resistance to UV light and ozone: UV light (a constituent of sunlight) and ozone attack unsaturation sites in the rubbers. These properties are important in rubber products exposed to sunlight and outer atmosphere during their service. Both these cause cracking and embrittlement of the rubbers. UV light starts degradation reactions which lead to excessive cross linking or depolymerisation. Ozone also decomposes the rubber to low molecular weight materials. Carbon blacks especially the smaller particle size grades, absorb UV light and protect the rubber. $BaSO_4$, $CaCO_3$ and TiO_2 (rutile) fillers can also absorb UV light. Aromatic oils, staining anti oxidants reduce resistance to UV light.

Ozone attacks rubbers when they are strained. Anti ozonants are often added to prolong the life of the products which may be exposed to ozone during their service. The role of physical and chemical anti ozonants are explained in the compounding ingredients chapter earlier. PVC is a polymeric anti ozonant for NBR. Blending EPDM into NR or SBR also improves their ozone resistance.

Low temperature resistance: Rubbers which crystallize on stressing like NR and CR can become stiff which is reversible. For NR, crystallisation-resistant grades have been developed while in CR, ester plasticizers should be avoided as they increase the rate of crystallization. Plasticisers which are compatible and of low freezing point are preferable.

Rubbers with lower Tgs should be preferred for best low temperature properties.

Electrical properties: They depend not only on the polarity of the rubber but also the filler grades and types.

Cure system:

Sulphur cure gives the best mechanical properties besides being the less expensive curative. For a compromise, in some applications, semi efficient cures (nearly equal sulphur and accelerator dosages) are chosen. Sulphurless cure also gives mainly monosulphide cross links.

Blending of elastomers:

Polymers due to the length of their molecular chains are mostly immiscible with each other. Blends show discrete areas of each elastomer varying from from 0.5 microns approx. for example, a successful blend of NR and SBR shows such a domain structure. The domain size depends on process conditions, viscosity of the polymers crystallisation etc.

The available rubbers do not have all the necessary combinations of properties and this makes blending imperative. Processing difficulties may be overcome in some cases while in others, cost may be reduced. Compounds, copolymers etc, can also be considered as blends but our definition will be confined only to polymer mixtures – rubber-rubber or rubber-plastics or plastic-plastic.

In this section we will consider rubber-rubber mixtures only.

Blending methods – the methods used commonly are latex blending, solution blending, powdered rubber mixing, mechanical blending, solution-latex blending etc.

Latex blending is fairly simple – this may give fine dispersion and can be commercially viable. However, for dry rubber processing, this method does not seem to work well.

Solution blending has limitations – one is the problem of the solvent-toxicity and flammability. Other problems are agglomeration into large domains during solvent evaporation after blending. Complete solvent removal is difficult.

Solution and latex mixing – this is for polymer pairs in which one is in a latex form (i.e. obtained by emulsion polymerization) and the other one is in solution form (solution polymerization).

In all the above methods, it is also possible to prepare carbon master batches too directly without mill mixing.

However, mechanical blending is the most viable method. High shear forces are needed for this blending which can be obtained in banbury mixers or 2-roll mills. For mill mixing, the viscosities of the rubbers must be brought to almost equal levels before blending. If cure rates of the rubbers vary, master batches may be prepared for both the rubbers separately before blending. For some other rubber pairs, addition of ingredients to the blend is a better method.

Mechano chemical methods are also mentioned in the literature. Mastication of a rubber itself is a mechano chemical process – thus co-masticating two rubbers in absence of oxygen leads to grafting of one rubber by the other.

Powdered rubber mixing is a viable method now a days as rubbers can be prepared from latex in powdered form by low cost methods. This can lead to easy blending.

Structural aspects of rubber blends: Many rubber pairs have been examined for homogeneity using various microscopic techniques and it has been found that very few pairs can be micro homogenous e.g. SBR-BR blends. This may appear surprising because from our basic knowledge about rubber chemistry, we should expect homogeneity in NR-BR blends and NR-SBR blends too.

Further, some blends of various SBRs varying in styrene contents were studied – even here, the blends were homogenous only if the difference in styrene content, between the polymers is less than 20%. Similarly, a mixture of high cis and low cis polybutadienes show heterogeneity at –40°C – this is because the high cis grade in the blend will crystallize at that temperature while the other one will no, leading to inhomogeneity.

Further, viscosities of the two polymers must be brought as close to each other as possible for achieving fine homogenous structure. A smaller nip gap has shown to ensure finer structure for NR-BR blending.

Another problem is, the mill/banbury mixing as done in the factories is time bound – equilibrium may not be reached within such a short time – thus instability of the fine structure cannot be ruled out.

The domain size can affect properties – finer structure in SBR-EPDM blends was shown to increase ozone resistance of the blend. This may be because the EPDM domains in the blend are closer to each other – it is the EPDM zones which contribute to ozone resistance.

Distribution of compounding ingredients: Carbon black, accelerators, anti oxidants etc. are attracted to different rubbers to different degrees. For example, carbon black is attracted to high unsaturation rubbers. Similarly, anti oxidants and accelerators being more polar will be attracted by the more polar rubber in a blend. All these can have a bearing on the properties of the blend.

If the curatives dissolve in one rubber to a greater extent, in a blend, then that rubber domain will have more of the curatives leading to grater state of cure of that rubber compared with the other one – thus there will be a cure mismatch and this will weaken the blend – the blend will be less homogenous as the more cured domains will be stronger than the other one – this will lead to failure at the weaker zones. This is aggravated by the difference in cure rates – thus in NR-IIR or NR-EPDM blends, the curatives will be more soluble in the NR and at the same time, NR will cure faster than IIR/EPDM – thus the blend will fail easily if we do not take steps to increase the state of cure in the IIR/EPDM phase.

In this connection, if the two rubbers can be made to cure independently of each other, we can get a viable blend. Example – a blend of NR and CR – here, CR can be cured by ZnO independently of the NR which is cured by sulphur and accelerator. We may recall that halogenation of IIR is done for the same reason.

For NR/IIR blends, if we can ensure better solubility of the accelerators in IIR, the blend performance can be improves – for this we can use longer chain thiurams or dithocarbamates – e.g. zinc di butyl (or higher alkyl) dithiocarbamates instead of ZDC – will have greater solubility in the less polar rubber - IIR or EPDM – this can lead to more uniformity in state of cure and better blend performance – better solubility of the accelerator in EPDM leads to faster cure in the EPDM phase and hence better performance. By grafting accelerators into EPDM too, blending of NR or SBR with EPDM will become easier.

Further, anti oxidants may migrate from one rubber to the other – if it happens from EPDM to SBR, the blend performance will improve – if it is in the other way, poor ageing resistance will result – as the anti oxidant is more needed in SBR – in this condition if it migrates from SBR to EPDM (EPDM can have longer life even without anti oxidant), optimum blend performance cannot be achieved.

Filler distribution: Carbon black will tend to migrate to a more unsaturated polymer – thus in NR-IIR blend, the carbon black will be more to the NR phase – it must be remembered that NR can give a high tensile strength without reinforcing filler while EPDM cannot. Thus the blend will have weak EPDM domains which can be starting points of mechanical failure – optimum performance cannot be achieved this way.

Similarly in a blend of NR and BR, carbon black tends to move to BR phase. This is due to the strain induced crystallization in NR which makes BR attract the carbon black. Among the various rubbers, carbon black affinity changes in the following order:

BR > SBR > CR > NBR > NR > EPDM > IIR

This means that the mixing procedure may have a bearing on the blend performance. Should we add all carbon black to BR or to NR or equally to both rubbers before blending the rubbers? Adding all carbon in BR and than blending NR may work better for the reasons explained above. The above mentioned order of affinities of rubbers to fillers may not be applicable for silica filler. This opens up new challenges as these days the use of dual filler-carbon in addition to silica in tyre tread formulations is on the increase.

Blending NR with BR increases the processing temperature of the blend and this improves properties. Adding liquid BR or liquid BR with COOH end groups to high cis BR improves the properties of the latter.

Blending high unsaturation grades of EPDM with SBR can lead to new applications for SBR.

The need for cost reduction in IIR and EPDM based products has led to extensive research on blending these rubbers with the low cost general purpose, diene rubbers.

The high heat, ozone and weather resistance of these rubbers is sought to be exploited in such blends. Problems due to the difference in cure rates of these rubbers with the general purpose rubbers are faced and many solutions have been found.

Die swell of a blend can be reduced if the black is located in the rubber with lower level of interaction with the carbon black. A blend of a saturated with unsaturated rubber shows a sheath-core configuration and plug flow. This is because the black goes to the unsaturated rubber and increases the viscosity of this component of the blend.

Increase in interfacial bonding between the rubbers in the blend leads to improvement in fatigue life–e.g. in a blend containing NR and BIIR, if a small amount of Butadiene-isoprene rubber (BIR) as a compatibiliser is added. Improvement of interfacial bonding also increases fatigue life of SBR-CIIR blend and tensile strength of EPDM – silicone blend. Compatibilisation will be explained later in this chapter. Adding a little CIIR to NR-BR blend reduced rolling resistance. This also reduces hysteresis without changing, the wet tack (i.e. stickiness when wet – this property is important while building, tyres, belts etc.)

The level of CIIR is high enough to cause good resilience at a low rate of deformation but causes high hysteresis at high rates – thus good balance between rolling and skid resistance.

To improve tack, a small amount of NR in a blend will help, provided NR migrates to the surface. A small amount of N-isopropoxy methacrylamide when added to a blend containing BR and a small amount of NR can cause cross linking of the former at cure temperatures and this leads to NR getting thrown up to the surface. CIIR on addition to CR improves tack of CR. 10% of CR in epichlorohydrin rubber – unsaturated rubber improves the interfacial bonding between the two rubbers. If the two rubbers do not have affinity to the black, the black agglomerates at interfaces and leads to better conductivity.

Applications of blends: Silicone rubbers blended with S–EB–S block copolymer give better bio compatibility. This is useful in medical products like adhesives.

CR is used in power transmission belts due to its ozone, oil and fatigue resistances. Its limitation is heat resistance which can be improved by blending with EPDM. This blend is compatibilised by adding a terpolymer containing ethylene-maleic anhydride and acrylic acid. The EPDM is grafted with maleic anhydride.

EPDM-NR blending can be improved by using chlorinated EPDM – the better ozone resistance can be exploited by this blend (the chlorine in the rubber compatibilises the blend). Blend of EPDM with NBR or SBR or poly sulphide can be improved by adding 1, 2 BR and CR to the blends. SBR-BR-EPDM blends can be successfully used without anti oxidants in tread compounds.

To compatibilise the IIR-NR blends, chlorination of the IIR is done. CIIR can easily blend with NR because the curing of the former is through ZnO while the latter is by sulphur – they proceed independent of each other and thus give a stable blend. In such blends the damping and low air permeability properties of IIR can be transmitted to NR. Blending EPDM to NBR give a good combination of solvent and heat resistance – this may be a low cost compromise over the use of HNBR which is very expensive. Blending BR into NBR improves abrasion resistance. Epichlorohydrin rubber-BR blends can give a good combination of low temperature resistance with low air permeability. Blending fluoro silicone rubber to silicone rubber leads to improvement in processibility, and compression set resistance besides cost reduction.

BR in NR leads to better abrasion resistance, higher processing temperatures, while the reverse leads to better tack, better processibility. Adding NR or SBR to BR in presence of salt of long chain fatty acid leads to better cutting resistance. Carboxyl terminated liquid BR improves, properties of BR. A small amount of liquid polymer with high Tg improves cut resistance of SBR. Brominated isobutylene-methyl styrene rubber in a blend of BIIR with or SBR or NR reduces air permeability and improves ageing, abrasion and fatigue resistance besides better adhesion to tyre carcass. This may lead to lighter tyres with lesser rolling resistance and low air permeability.

FKMs are known for high chemical resistance but amine additives in high performance engine oils may harden the rubber when used as seals (leading to cracking). FKMs also have poor resistance to low temperatures. ACMs may resist this fuel and low temperatures but have poor heat resistance. A blend of these two rubbers may solve this problem – this can be achieved by choosing the proper grades of these rubbers – the FKM should be curable by bis phenol A while the ACM should be curable by amine or sulphur-soap system. Compatibility should be optimum-too low leads to easy mechanical failure while too high may lead to considerable reduction in Tg. Another advantage is the low cost of this blend compared with the corresponding HNBR based product.

The major uses of polymer blending are found in all rubber products like tyres, belting, hoses, other moulded goods.

Compatibilisation of blends: The role of a compatibiliser is to reduce the size of the interface and adhesion between the polymers or to reduce the size of the dispersed phase (often the minor component of the blend) and stabilize the dispersion against agglomeration during processing and service of the product.

Compatibilisation is of two types – physical and chemical. Physical compatibilisers are often block and graft copolymers – they act in a way similar to that of a surfactant – the more polar component of the blend gets attracted to the more polar part of the compatibiliser while the less polar polymer to the less polar part.

In chemical compatibilisation, often maleic anhydride is grafted onto a polymer. The anhydride attracts the polar polymer and may even react with it as in case of EPDM grafted with maleic anhydride blending with nylon. The anhydride reacts with nylon and compatibilises the blend. Examples of this concept is also explained in TPE chapter where blends of PP with NBR or EPDM has been explained. We may also have ethylene-acrylic acid ionomer or other such systems as compatibilisers.

Compatibilisation by surface activation – e.g. scrap tyre can be surface modified and blended with other polymers to get low cost blends with better performance. Another example is scrap EPDM which can be treated with a mixture of SO_2, F_2 and N_2 gases, can be blended with TPU to give low cost products.

Compatibilisation by cross linking – the two phases which are separated can be bound chemically using cross linking between the phases – as in dynamic cross linking for preparing blend type TPE. Often in the interface, block or graft copolymers are formed – in many blends evidence for this is seen. The role of liquid BR as a compatibiliser seems to be through this route besides the plasticization of the phases by the liquid BR. Liquid BR also acts as a coagent for blending EPDM with SBR –.here too compatibilisation is through cross linking. Many blends are possible mainly due to compatibilisation by cross linking or graft or block copolymer formations in situ.

Bibliography

1. Brydson J.A., Rubber Chemistry, Applied Science Publishers, 1978.
2. Franta I, Elastomers and Rubber Compounding Materials, Elsevier, 1989.
3. Morton M, Rubber Technology, Springer, 1995.
4. Mark, J.E., Erman B, Eirch F.R., The Science and Technology of Rubbers, Elsevier Acad, Press, 2005.

Index